Analog Circuits Cookbook

Analog Circuits Cookbook

Second edition

Ian Hickman BSc (Hons), CEng, MIEE, MIEEE

Newnes

OXFORD AUCKLAND BOSTON JOHANNESBURG MELBOURNE NEW DELHI

Newnes
An imprint of Butterworth-Heinemann
Linacre House, Jordan Hill, Oxford OX2 8DP
225 Wildwood Avenue, Woburn, MA 01801-2041
A division of Reed Educational and Professional Publishing Ltd

A member of the Reed Elsevier plc group

First published 1995
Second edition 1999

© Ian Hickman 1995, 1999

All rights reserved. No part of this publication may be reproduced in any material form (including photocopying or storing in any medium by electronic means and whether or not transiently or incidentally to some other use of this publication) without the written permission of the copyright holder except in accordance with the provisions of the Copyright, Designs and Patents Act 1988 or under the terms of a licence issued by the Copyright Licensing Agency Ltd, 90 Tottenham Court Road, London, England W1P 9HE. Applications for the copyright holder's written permission to reproduce any part of this publication should be addressed to the publishers

British Library Cataloguing in Publication Data
A catalogue record for this book is available from the British Library.

ISBN 0 7506 4234 3

Library of Congress Cataloguing in Publication Data
A catalogue record for this book is available from the Library of Congress.

Typeset by Tek-Art, Croydon, Surrey
Printed and bound in Great Britain by
Biddles Ltd, Guildford and King's Lynn

Contents

Preface to second edition   ix

1 Advanced circuit techniques, components and concepts   **1**
Negative approach to positive thinking   1
March 1993, pages 258–261
Logamps for radar – and much more   10
April 1993, pages 314–317
Working with avalanche transistors   16
March 1996, pages 219–222
Filters using negative resistance   26
March 1997, pages 217–221
Big surprises ... in small packages   39
May 1997, pages 371–376, 440

2 Audio   **57**
Low distortion audio frequency oscillators   57
April 1992, pages 345–346
Notes on free phasing   61
February 1996, pages 124–128
Music in mind   73
October 1996, pages 730–734
Filter variations   84
October 1996, pages 769–772
Camcorder dubber   94
September 1997, pages 730–731

3 Measurements (audio and video) — 99

Four opamp inputs are better than two — 99
May 1992, pages 399–401

DC accurate filter plays anti-alias role — 104
June 1992, pages 497–499

Bootstrap base to bridge building — 110
October 1992, pages 868–870

Mighty filter power in minuscule packages — 116
May 1993, pages 399–403

'Scope probes – active and passive — 126
May 1996, pages 366–372

4 Measurements (rf) — 142

Measuring detectors (Part 1) — 142
November 1991, pages 976–978

Measuring detectors (Part 2) — 147
December 1991, pages 1024–1025

Measuring L and C at frequency – and on a budget — 151
June 1993, pages 481–483

Add on a spectrum analyser — 160
December 1993, pages 982–989

Wideband isolator — 177
March 1998, pages 214–219

5 Opto — 191

Sensing the position — 191
November 1992, pages 955–957

Bringing the optoisolator into line — 198
December 1992, pages 1050–1052

Light update — 205
September 1996, pages 674–679

A look at light — 213
June 1997, pages 466–471

6 Power supplies and devices — 228

Battery-powered instruments — 228
February 1981, pages 57–61

The MOS controlled thyristor — 242
September 1993, pages 763–766

Designer's power supply — 252
January 1997, pages 26–32

7 RF circuits and techniques — **268**

Homodyne reception of FM signals — 268
November 1990, pages 962–967

LTPs and active double balanced mixers — 281
February 1993, pages 126–128

Low power radio links — 288
February 1993, pages 140–144

Noise — 302
February 1998, pages 146–151

Understanding phase noise — 316
August 1997, pages 642–646

Index — **329**

Preface to second edition

Electronics World + *Wireless World* is undoubtedly the foremost electronics magazine in the UK, being widely read by both professional electronics engineers on the one hand and electronics hobbyists and enthusiasts on the other, in the UK, abroad and indeed around the world. The first article of mine to feature in the magazine, then called simply *Wireless World,* appeared back in the very early 1970s. Or was it the late 1960s; I can't remember. Since then I have become a more frequent – and latterly a regular – contributor, with both the 'Design Brief' feature and occasional longer articles and series. With their straightforward non-mathematical approach to explaining modern electronic circuit design, component applications and techniques, these have created some interest and the suggestion that a collection of them might appear in book form found general approval among some of my peers in the profession. The first edition of this book was the result. A sequel, *Hickman's Analog and R.F. Circuits*, containing a further selection of articles published in *Electronics World* (as it is now known), was published subsequently.

Since the appearance of the first edition of the *Analog Circuits Cookbook* in 1995, a lot of water has flowed under the bridge, in technical terms. Some of the articles it contains are thus no longer so up-to-the-minute, whilst others are still entirely relevant and very well worth retaining. So this second edition of the *Analog Circuits Cookbook* has been prepared, retaining roughly half of the articles which appeared in the first edition, and replacing the rest with other articles which have appeared more recently in *Electronics World*.

Inevitably, in the preparation for publication of a magazine which appears every month, the occasional 'typo' crept into the articles as published, whilst the editorial exigencies of adjusting an article to fit the space available led to the occasional pruning of the text. The opportunity has been taken here of restoring any excised material and of correcting all (it is hoped) errors in the articles as they appeared in the magazine. The articles have been gathered together in chapters under subject headings, enabling readers to home in rapidly on any area in which they are particularly interested. A brief introduction has also been added to each, indicating the contents and the general drift of the article.

1 Advanced circuit techniques, components and concepts

> **Negative components**
>
> Negative components may not be called for every day, but can be extremely useful in certain circumstances. They can be easily simulated with passive components plus opamps and one should be aware of the possibilities they offer.

Negative approach to positive thinking

There is often felt to be something odd about negative components, such as negative resistance or inductance, an arcane aura setting them apart from the real world of practical circuit design. The circuit designer in the development labs of a large firm can go along to stores and draw a dozen 100 kΩ resistors or half a dozen 10 µF tantalums for example, but however handy it would be, it is not possible to go and draw a −4.7 kΩ resistor. Yet negative resistors would be so useful in a number of applications; for example when using mismatch pads to bridge the interfaces between two systems with different characteristic impedances. Even when the difference is not very great, for example testing a 75 Ω bandpass filter using a 50 Ω network analyser, the loss associated with each pad is round 6 dB, immediately cutting 12 dB off how far down you can measure in the stopband. With a few negative resistors in the junk box, you could make a pair of mismatch pads with 0 dB insertion loss each.

But in circuit design, negative component values do turn up from time to time and the experienced designer knows when to accommodate them, and when to redesign in order to avoid them. For example, in a filter design it may turn out that a −3 pF capacitor, say, must be added between nodes X and Y. Provided that an earlier stage of the computation has resulted in a capacitance of more than this value appearing between those nodes, there is no problem; it is simply reduced by 3 pF to give the final value. In the case where the final value is still negative, it may be necessary to redesign to avoid the problem, particularly at UHF and above. At lower frequencies, there is always the option of using a 'real' negative capacitor (or something that behaves exactly like one); this is easily implemented with an 'ordinary' (positive) capacitor and an opamp or two, as are negative resistors and inductors. However, before looking at negative components using active devices, note that they can be implemented in entirely passive circuits if you know how (Roddam, 1959). Figure 1.1(a) shows a parallel tuned circuit placed in series with a signal path, to act as a trap, notch or rejector circuit. Clearly it only works

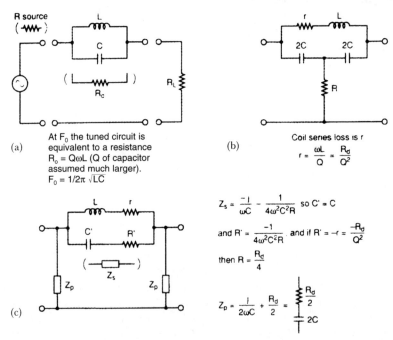

Figure 1.1 (a) A parallel tuned circuit used as a rejector. The notch depth is set by the ratio of the tuned circuit's dynamic resistance R_d and the load resistance R_l. At F_0 the tuned circuit is equivalent to a resistance $R_d = Q\omega L$ (Q of capacitor assumed much larger). $F_0 = 1/2\pi \sqrt{(LC)}$. (b) The circuit modified to provide a deep notch, tuned frequency unchanged. Coil series losses $r = \omega L/Q = R_d/Q^2$. (c) As (b) but with the star network transformed to the equivalent delta network. $Z_s = (-j/\omega C) - 1/(4\omega^2 C^2 R)$. So $C' = C$ and $R' = -1/(4\omega^2 C^2 R)$ and if $R' = -r = -R_d/Q^2$ then $R = R_d/4$, $Z_p = (j/2\omega C) + (R_d/2)$

well if the load resistance R_l is low compared with the tuned circuit's dynamic impedance R_d. If R_l is near infinite, the trap makes no difference, so R_d should be much greater than R_l; indeed, ideally we would make R_d infinite by using an inductor (and capacitor) with infinite Q. An equally effective ploy would be to connect a resistance of $-R_d$ in parallel with the capacitor, cancelling out the coil's loss exactly and effectively raising Q to infinity. This is quite easily done, as in Figure 1.1(b), where the capacitor has been split in two, and the tuned circuit's dynamic resistance R_d ($R_d = Q\omega L$, assuming the capacitor is perfect) replaced by an equivalent series loss component r associated with the coil ($r = \omega L/Q$). From the junction of the two capacitors, a resistor R has been connected to ground. This forms a star network with the two capacitors, and the next step is to transform it to a delta network, using the star-delta equivalence formulae. The result is as in Figure 1.1(c) and the circuit can now provide a deep notch even if R_l is infinite, owing to the presence of the shunt impedance Z_p across the output, if the right value for R is chosen. So, let $R' = -r$, making the resistive component of Z_s (in parallel form) equal to $-R_d$. Now R' turns out to be $-1/(4\omega^2 C^2 R)$ and equating this to $-r$ gives $R = R_d/4$.

Negative inductor

Now for a negative inductor, and all entirely passive – not an opamp in sight. Figure 1.2(a) shows a section of constant-K lowpass filter acting as a lumped passive delay line. It provides a group delay $dB/d\omega$ of $\sqrt{(LC)}$ seconds per section. Figure 1.2(b) at dc and low frequencies, maintained fairly constant over much of the passband of the filter. A constant *group* delay (also known as envelope delay) means that all frequency components passing through the delay line (or through a filter of any sort) emerge at the same time as each other at the far end, implying that the *phase* delay $B = \omega \sqrt{(LC)}$ radians per section is proportional to frequency. (Thus a complex waveform such as an AM signal with 100% modulation will emerge unscathed, with its envelope delayed but otherwise preserved unchanged. Similarly, a squarewave will be undistorted provided all the significant harmonics lie within the range of frequencies for which a filter exhibits a constant group delay. Constant group delay is thus particularly important for an IF bandpass filter handling phase modulated signals.) If you connect an inductance L' (of suitable value) in series with each of the shunt capacitors, the line becomes an 'm-derived' lowpass filter instead of a constant-K filter, with the result that the increase of attenuation beyond the cut-off frequency is much more rapid. However, that is no great benefit in this

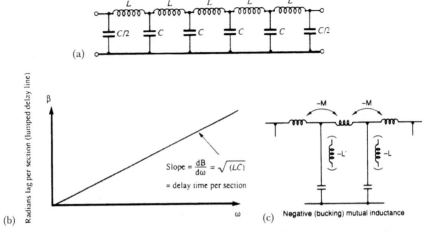

Figure 1.2 (a) Basic delay line – (b) providing a delay of √(LC) seconds per section at dc and low frequencies. (c) Connection of negative inductance in the shunt arms to linearise the group delay over a larger proportion of the filter's passband. Not a physical component, it is implemented by negative mutual inductance (bucking coupling) between sections of series inductance

application, a delay line is desired above all to provide a constant group delay over a given bandwidth and the variation in group delay of an *m*-derived filter is much worse even than that of a constant-*K* type. Note that L' may not be a separate physical component at all, but due to mutual coupling between adjacent sections of series inductance, often wound one after the other, between tapping points on a cylindrical former in one long continuous winding. If the presence of shunt inductive components L' makes matters worse than the constant-*K* case, the addition of negative L' improves matters. This is easily arranged (Figure 1.2(c)) by winding each series section of inductance in the opposite sense to the previous one.

Real pictures

Now for some negative components that may, in a sense, seem more real, implemented using active circuitry. Imagine connecting the output of an adjustable power supply to a 1 Ω resistor whose other end, like that of the supply's return lead, is connected to ground. Then for every volt positive (or negative) that you apply to the resistor, 1 A will flow into (or out of) it. Now imagine that, without changing the supply's connections, you arrange that the previously earthy end of the resistor is automatically jacked up to twice the power supply output voltage, whatever that happens to be. Now, the voltage across the resistor is always equal to the power supply output voltage, but of the opposite polarity. So when, previously, current flowed into the resistor, it now supplies an output current, and vice versa. With the current always of the wrong sign, Ohm's law will still hold if we label the value of the resistor −1 Ω. Figure 1.3(a) shows the scheme, this time put to use to provide a capacitance of −C μF, and clearly substituting L for C will give a negative inductance. For a constant applied ac voltage, a negative inductance will draw a current leading by 90° like a capacitor, rather than lagging like a positive inductor. But like a positive inductor, its impedance will still rise with frequency. Figure 1.3 also

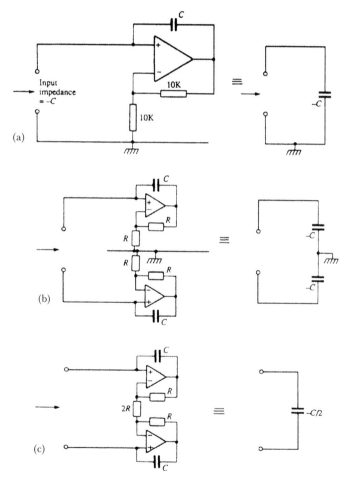

Figure 1.3 (a) Unbalanced negative capacitor (one end grounded). (b) Balanced, centre grounded negative capacitor. (c) Floating negative capacitor

shows how a negative component can be balanced, or even floating. It will be clear that, if in Figure 1.3(a), C is 99 pF and the circuit is connected in parallel with a 100 pF capacitor, 99% of the current that would have been drawn from an ac source in parallel with the 100 pF capacitor will now be supplied by the opamp via C, leaving the source 'seeing' only 1 pF. Equally, if the circuit is connected in parallel with an impedance which, at some frequency, is higher than the reactance of C, the circuit will oscillate; this circuit is 'short circuit stable'.

Negative capacitance

A negative capacitance can be used to exterminate an unwanted positive capacitance, which can be very useful in certain applications where stray capacitance is deleterious to performance yet unavoidable. A good example is the N-path (commutating) bandpass filter which, far from being an academic curiosity, has been used both in commercial applications, such as FSK modems for the HF band, and in military applications. One disadvantage of this type of bandpass filter is that the output waveform is a fairly crude, N-step approximation to the input, N being typically 4, requiring a good post filter to clean things up. But on the other hand, it offers exceptional values of Q. Figure 1.4(a) illustrates the basic scheme, using a first-order section. If a sinusoidal input at exactly a quarter of the clock frequency is applied at v_i (Figure 1.4(a)), so that the right-hand switch closes for a quarter of a cycle, spanning the negative peak of the input, and the switch second from left acts similarly on the positive peak, the capacitors will charge up so that v_o is a stepwise approximation to a sinewave as in Figure 1.4(b), bottom left. The time constant will not be CR but $4CR$, since each capacitor is connected via the resistor to the input for only 25% of the time. If the frequency of the input sinewave differs from $F_{clock}/4$ (either above or below) by an amount less than $1/(2\pi 4CR)$, the filter will be able to pass it, but if the frequency offset is greater, then the output will be attenuated, as shown in Figure 1.4(c). Depending upon the devices used to implement the filter, particularly the switches, F_{clock} could be as high as tens of kHz, whereas C and R could be as large as 10 µF and 10 MΩ, giving (in principle) a Q of over 10 million.

Kundert filter

The same scheme can be applied to a Kundert filter section, giving a four pole bandpass (two pole LPE – low pass equivalent) section (Figure 1.4(c) and (d)). Figure 1.5(a) shows the response of a five

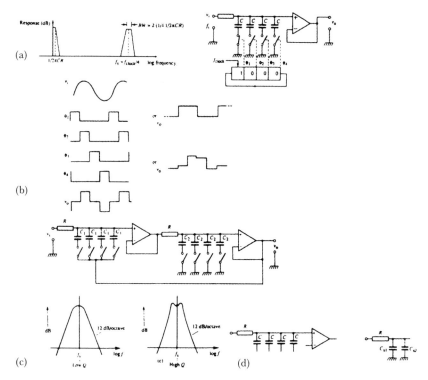

Figure 1.4 (a) One pole lowpass equivalent (LPE) N-path bandpass filter section. A solitary 1 circulating in a shift register is ony one of the many ways of producing the four-phase drive waveform shown in (b). (b) Waveforms associated with (a). The exact shape of v_o when $f_i = F_{clock}/4$ exactly will depend on the relative phasing of v_i and the clock waveform. For very small difference between f_i and $F_{clock}/4$ the output will continuously cycle between the forms shown and all intermediate shapes. (c) Second-order N-path filter, showing circuit frequency response. $Q = 1/\sqrt{(C_1/C_2)}$, exactly as for the lowpass case. (d) Stray capacitance. Showing the stray capacitance to ground, consisting of opamp input capacitance C_{s2} plus circuit and component capacitance to ground with all switches open at C_{s1}

pole LPE 0.5 dB ripple Chebychev N-path filter based on a Sallen and Key lowpass prototype, with a 100 Hz bandwidth centred on 5 kHz. The 6 to 60 dB shape factor is well under 3:1 with an ultimate rejection of well over 80 dB. However, the weak point in this type of filter is stray capacitance across each group of switched capacitors. This causes the 'smearing' of charge from one capacitor into the next, which has the unfortunate effect in high Q second-order sections of lowering the frequency of the two peaks slightly and also of unbalancing their amplitude. The higher the centre frequency, the

8 Analog circuits cookbook

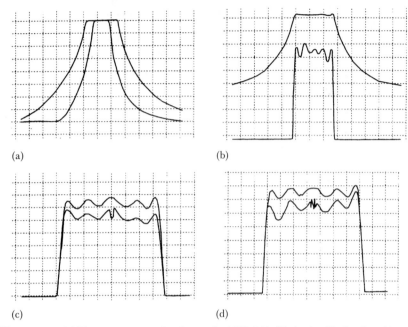

Figure 1.5 (a) The response of a five pole LPE 0.5 dB ripple Chebychev N-path filter based on a Salen and Key lowpass prototype, with a 100 Hz bandwidth centred on 5 kHz, 10 dB/div. vertical, 50 Hz and 100 Hz/div. horizontal. (At a 20 kHz centre frequency, its performance was grossly degraded.) (b) A five pole LPE Chebychev N-path filter with a 100 Hz bandwidth centred on 20 kHz, using the Kundert circuit for the two pole stage, and its response (10 dB and 1 dB/div. vertical, 50 Hz/div. horizontal). (c) The passband of (b) in more detail, with (upper trace) and without –39 pF to ground from point C. 1 dB/div. vertical; 20 Hz per div. horizontal. Note: the gain was unchanged; the traces have been separated vertically for clarity. (d) The passband of (b) in more detail, with –39 pF (upper trace) and with –100 pF to ground from point C; overcompensation reverses the slope

smaller the value of the switched capacitors, the narrower the bandwidth or the higher the section Q, the more pronounced is the effect. This results in a crowding together of the peaks of the response on the higher frequency side of the passband and a spreading of them further apart on the lower, producing a slope up across the passband (Figure 1.5(a)), amounting in this case to 1 dB. Increasing the clock frequency to give a 20 kHz centre frequency results in a severely degraded passband shape, due to the effect mentioned. Changing the second-order stage to the Kundert circuit (Figure 1.5(b)) improves matters by permitting the use of larger capacitors; C_2 can be as large as C_1 in the Kundert circuit, whereas in the Salen and Key circuit, the ratio is defined by the desired stage Q. With this modification, the filter's response is as in Figure 1.5(b). The modification restores the correct response of the high Q two pole output section, but the downward shift of the peaks provided by the three pole input section results in a downward overall passband slope with increasing frequency. Note the absence of any pip in the centre of the passband due to switching frequency breakthrough. (If the charge injection via each of the switches was identical, there would be no centre frequency component, only a component at four times the centre frequency, i.e. at the switching frequency. Special measures, not described here, are available to reduce the switching frequency breakthrough. Without these, the usable dynamic range of an N-path filter may be limited to as little as 40 dB or less; with them the breakthrough was reduced to –90 dBV. Figure 1.5(b) was recorded after the adjustment of the said measures.) The slope across the passband is shown in greater detail in Figure 1.5(c) (lower trace) – this was recorded before the adjustment, the centre frequency breakthrough providing a convenient 'birdie marker' indicating the exact centre of the passband. The upper trace shows the result of connecting –39 pF to ground from point C_2 of Figure 1.5(b), correcting the slope. Figure 1.5(d) shows the corrected passband (upper trace) and the effect of increasing the negative capacitance to –100 pF (lower trace), resulting in overcompensation.

These, and other examples which could be cited, show the usefulness of negative components to the professional circuit designer. While they may not be called for every day, they should certainly be regarded as a standard part of the armoury of useful techniques.

Acknowledgements

Figures 1.2(a), (b), 1.3 and 1.4 are reproduced with permission from Hickman, I. (1990) *Analog Electronics*, Heinemann Newnes, Oxford.

References

Hickman, I. (1993) CFBOs: delivering speed at any gain? *Electronics World + Wireless World*, January, 78–80.
Roddam, T. (1959) The Bifilar-T circuit. *Wireless World*, February, 66–71.

Logarithmic amplifiers

Logarithmic amplifiers (logamps for short) have long been employed in radar receivers, where log IF strips were made up of several or many cascaded log stages. Now, logamps with dynamic ranges of 60, 70 or even 80 dB are available in a single IC, and prove to have a surprisingly wide range of applications.

Logamps for radar – and much more

The principles of radar are well known: a pulse of RF radiation is transmitted from an antenna and the echo – from, for example, an aeroplane – is received by (usually) the same antenna, which is generally directional. In practice, the radar designer faces a number of problems; for example, in the usual single antenna radar, some kind of a T/R switch is needed to route the Transmit power to the antenna whilst protecting the Receiver from overload, and at other times routeing all of the minuscule received signal from the antenna to the receiver. From then on, the problem is to extract wanted target returns from clutter (background returns from clouds, the ground or sea, etc.) or, at maximum range, receiver noise, in order to maximise the Probability of Detection P_d whilst minimising the Probability of False Alarm P_{fa}.

With the free-space inverse square law applying to propagation in both the outgoing and return signal paths, the returned signal power from a given sized target is inversely proportional to the fourth power of distance: the well-known basic R^4 radar range law. With the consequent huge variations in the size of target returns with range, a fixed gain IF amplifier would be useless. The return from a target at short range would overload it, whilst at long range the signal would be too small to operate the detector. One alternative is a swept gain IF amplifier, where the gain is at minimum immediately following the transmitted pulse and increases progressively with elapsed time thereafter, but this scheme has its own difficulties and is not always convenient. A popular arrangement, therefore, is the logarithmic amplifier. Now, if a target flies towards the radar, instead of the return signal rising 12 dB for each halving of the range, it increases by a fixed increment, determined by the scaling of the amplifier's log law.

This requires a certain amount of circuit ingenuity, the basic arrangement being an amplifier with a modest, fixed amount of gain, and ability to accept an input as large as its output when overdriven. Figure 1.6 explains the principle of operation of a true log amplifier

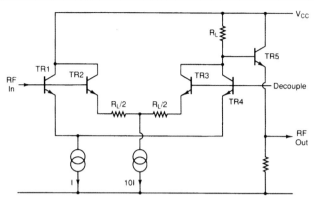

Figure 1.6 *True log amplifier. At low signal levels, considerable gain is provided by Tr_1 and Tr_4, which have no emitter degeneration (gain setting) resistors. At higher levels, these transistors limit, but the input is now large enough to cause a significant contribution from Tr_2 and Tr_3, which operate at unity gain. At even larger signal levels, these also limit, so the gain falls still further. At very low input signal levels, the output from the stage starts to rise significantly, just before a similar preceding stage reaches limiting*

stage, such as the GEC Plessey Semiconductors *SL531*. An IF strip consisting of a cascade of such stages provides maximum gain when none of the stages is limiting. As the input increases, more and more stages go into limiting, starting with the last stage, until the gain of the whole strip falls to ×1 (0 dB). If the output of each stage is fitted with a diode detector, the sum of the detected output voltages will increase as the logarithm of the strip's input signal. Thus a dynamic range of many tens of dB can be compressed to a manageable range of as many equal voltage increments.

A strip of true logamps provides, at the output of the last stage, an IF signal output which is hard limited for all except the very smallest inputs. It thus acts like the IF strip in an FM receiver, and any phase information carried by the returns can be extracted. However, the 'amplitude' of the return is indicated by the detected (video) output; clearly if it is well above the surrounding voltage level due to clutter, the target can be detected with high P_d and low P_{fa}. Many (in fact most) logamps have a built-in detector: if the logamp integrates several stages, the detected outputs are combined into a single video output. If target detection is the only required function, then the limited IF output from the back end of the strip is in fact superfluous, but many logamps make it available anyway for use if required. The GEC Plessey Semiconductors *SL521* and *SL523* are single and two stage logamps with bandwidths of 140 MHz and 100 MHz respectively, the two detected outputs in the *SL523* being combined internally into a single video output. These devices may be simply cascaded, RF output of one to the RF input of the next, to provide log ranges of 80 dB or more. The later *SL522*, designed for use in the 100–600 MHz range, is a successive detection 500 MHz 75 dB log range device in a 28 pin package, integrating seven stages and providing an on-chip video amplifier with facilities for gain and offset adjustment, as well as limited IF output.

The design of many logamps, such as those just mentioned, see GEC Plessey Semiconductors Professional Products I.C. Handbook, includes internal on-chip decoupling capacitors which limit the lower frequency of operation. These are not accessible at package pins and so it is not possible to extend the operating range down to lower frequencies by strapping in additional off-chip capacitors. This limitation does not apply to the recently released Analog Devices *AD606*, which is a nine stage 80 dB range successive detection logamp with final stage providing a limited IF output. It is usable to beyond 50 MHz and operates over an input range of –75 dBm to +5 dBm. The block diagram is shown in Figure 1.7(a), which indicates the seven cascaded amplifier/video detector stages in the main signal path preceding the final limiter stage, and a further two amplifier/video detector 'lift' stages (high-end detectors) in a side-chain fed via a 22 dB attenuator. This extends the operational input range above the level at which the main IF cascade is limiting solidly in all stages. Pins 3 and 4 are normally left open circuit, whilst OPCM (output common, pin 7) should be connected to ground. The 2 μA per dB out of the one pole filter, flowing into the 9.375 kΩ resistor between pins 4 and 7 (ground) defines a log slope law of 18.75 mV/dB at the input to the ×2 buffer amplifier input (pin 5) and hence of 37.5 mV/dB (typically at 10.7 MHz) at the video output VLOG, pin 6. The absence of any dependence on internal coupling or decoupling capacitors in the main signal path means that the device operates in principle down to dc, and in practice down to 100 Hz or less (Figure 1.7(b)). In radar applications, the log law (slope) and intercept (output voltage with zero IF input signal level) are important. These may be adjusted by injecting currents derived from VLOG and from a fixed reference voltage respectively, into pin 5. A limited version of the IF signal may be taken from LMLO and/or LMHI (pins 8 and 9, if they are connected to the +5 V supply rail via 200 Ω resistors), useful in applications where information can be obtained from the phase of the IF output. For this purpose, the variation of phase with input signal level is specified in the data sheet. If an IF output is not required, these pins should be connected directly to +5 V.

The wide operating frequency range gives the chip great versatility. For example, in an FM receiver the detected video output with its logarithmic characteristic makes an ideal RSSI (received signal

Advanced circuit techniques, components and concepts 13

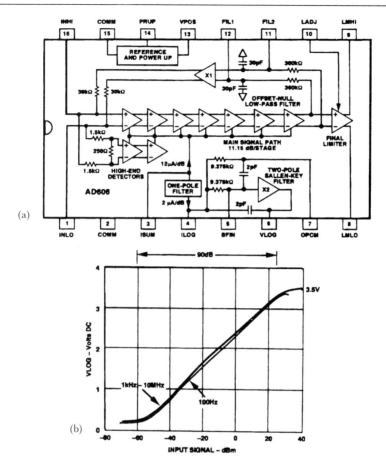

Figure 1.7 (a) Block diagram of the Analog Devices AD606 50 MHz, 80 dB demodulating logarithmic amplifier with limiter output; (b) shows that the device operates at frequencies down to the audio range

strength indicator). It can also be used in a low cost RF power meter and even in an audio level meter. To see just how this would work, the device can be connected as in Figure 1.8(a), which calls for a little explanation. Each of the detectors in the log stages acts as a full-wave rectifier. This is fine at high input signal levels, but at very low levels the offset in the first stage would unbalance the two half cycles: indeed, the offset could be greater than the peak-to-peak input swing, resulting in no rectification at all. Therefore, the device includes an internal offset-nulling servo-loop, from the output of the penultimate stage back to the input stage. For this to be effective at dc the input must be ac coupled as shown and, further, the input should present a low impedance at INLO and INHI (pins 1 and 16) so that the input

14 *Analog circuits cookbook*

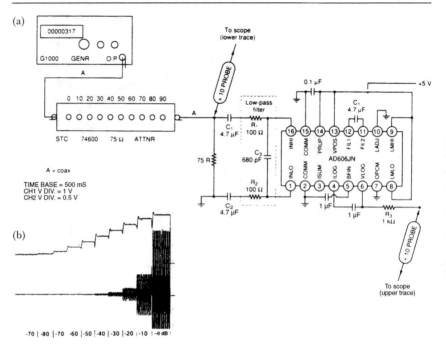

Figure 1.8 *(a) Circuit used to view the log operation at low frequency; (b) input signal (lower trace), increasing in 10 dB steps and the corresponding VLOG output (upper trace). The dip at the end of each 10 dB step is due to the momentary interruption of the signal as the attenuator setting is reduced by 10 dB and the following overshoot to the settling of the Sallen and Key filter*

stage 'sees' only the ac input signal and not any ac via the nulling loop. Clearly the cut-off frequency of the internal Sallen and Key lowpass filter driving the VLOG output is high, so that, at audio, the log output at pin 6 will slow a rather squashed looking full-wave rectified sinewave. This is fine if the indicating instrument is a moving coil meter, since its inertia will do the necessary smoothing. Likewise, many DVMs incorporate a filter with a low cut-off frequency on the dc voltage ranges. However, as it was intended to display VLOG on an oscilloscope, the smoothing was done in the device itself. The cut-off frequency of the Sallen and Key filter was lowered by bridging 1 μF capacitors across the internal 2 pF capacitors, all the necessary circuit nodes being available at the device's pins. The 317 Hz input to the chip and the VLOG output where displayed on the lower and upper traces of the oscilloscope respectively (Figure 1.8(b)). With the attenuator set to 90 dB, the input was of course too small to see. The attenuation was reduced to zero in 10 steps, all the steps being clearly visible on the upper trace. The 80 to 70 dB step is somewhat

Advanced circuit techniques, components and concepts 15

compressed, probably owing to pick-up of stray RF signals, since the device was mounted on an experimenter's plug board and not enclosed in a screened box. With its high gain and wide frequency response, this chip will pick up any signals that are around.

The device proved remarkably stable and easy to use, although it must be borne in mind that pins 8 and 9 were connected directly to the decoupled positive supply rail, as the limited IF output was not required in this instance.

Figure 1.9(a) shows how a very simple RF power meter, reading directly in dBm, can be designed using this IC. Note that here, the

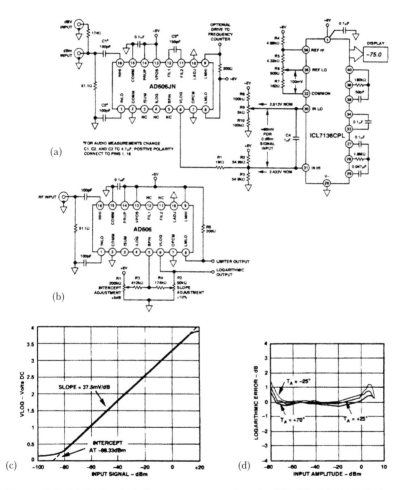

Figure 1.9 (a) A simple RF power meter using the AD606; (b) AD606 slope and intercept adjustment using pin 5; (c) AD606 nominal transfer function; (d) AD606 log conformance at 10.7 MHz

slope and intercept adjustment have been implemented externally in the meter circuit, rather than internally via pin 5. Where this is not possible, the arrangement of Figure 1.9(b) should be used.

This is altogether a most useful device: if it is hung on the output of a TV tuner with a sawtooth on its varactor tuning input, it provides a simple spectrum analyser with log display. Clearly, though, some extra IF selectivity in front of the *AD606* would be advisable. The later AD8307 operates to 500 MHz.

Acknowledgements

Figures 1.7(a), (b), 1.8 and 1.9 are reproduced with permission from *EW + WW*, April 1993, 314–317.

Avalanche transistor circuits

I was glad of the opportunity to experiment with some intriguing devices with rather special properties. Rather neglected until recently, new applications have rekindled interest in avalanche transistors.

Working with avalanche transistors

Introduction

I have been fascinated by avalanche transistor circuits ever since I first encountered them in the early 1960s. They have probably been known since the earliest days of silicon transistors but I have never heard of them being implemented with germanium devices, though some readers may know otherwise. One important use for them was in creating extremely fast, narrow pulses to drive the sampling gate in a sampling oscilloscope. Such oscilloscopes provided, in the late 1950s, the then incredible bandwidth of 2 GHz, at a time when other oscilloscopes were struggling, with distributed amplifiers and special cathode ray tubes, to make a bandwidth of 85 MHz. Admittedly those early sampling oscilloscopes were plagued by possible aliased responses and, inconveniently, needed a separate external trigger, but they were steadily developed over the years, providing, by the 1970s, a bandwidth of 10–14 GHz. The latest digital sampling oscilloscopes provide bandwidths of up to 50 GHz, although like their analog predecessors they are limited to displaying repetitive waveforms, making them inappropriate for some of the more difficult oscilloscope applications, such as glitch capture.

The basic avalanche transistor circuit is very simple, and a version published in the late 1970s (Ref. 1) apparently produced a 1 Mpulse/sec pulse train with a peak amplitude of 11 V, a half-amplitude pulse width of 250 ps and a risetime of 130 ps. This with a *2N2369*, an unremarkable switching transistor with a 500 MHz f_t and a C_{obo} of 4 pF. The waveform, reproduced in the article, was naturally captured on a sampling oscilloscope.

The avalanche circuit revisited

Interest in avalanche circuits seems to have flagged a little after the 1970s, or perhaps it is that the limited number of specialised uses for which they are appropriate resulted in the spotlight always resting elsewhere. Another problem is the absence of transistor types specifically designed and characterised for this application. But this situation has recently changed, due to the interest in high-power laser diodes capable of producing extremely narrow pulses for ranging and other purposes, in Pockel cell drivers, and in streak cameras, etc. Two transistors specifically characterised for avalanche pulse operation, types *ZTX413* and *ZTX415* (Ref. 2), have recently appeared, together with an application note (Ref. 3) for the latter.

The avalanche transistor depends for its operation on the negative resistance characteristic at the collector. When the collector voltage exceeds a certain level, somewhere between V_{ceo} and V_{cbo}, depending on the circuit configuration, the voltage gradient in the collector region exceeds the sustainable field strength, and hole–electron pairs are liberated. These are accelerated by the field, liberating others in their turn and the current thus rises rapidly, even though the voltage across the device is falling. The resultant 'plasma' of carriers results in the device becoming almost a short circuit, and it will be destroyed if the available energy is not limited. If the current in the avalanche mode, I_{USB}, and the time for which it is allowed to flow are controlled, then reliable operation of the device can be ensured, as indicated in Figure 1.10 for the *ZTX415*. From this it can be seen that for 50 ns wide pulses, a pulse current of 20 A can be passed for an indefinite number of pulses without device failure, provided of course that the duty cycle is kept low enough to remain well within the device's 680 mW allowable average total power dissipation $P_{tot.}$

Figure 1.11 shows a simple high-current avalanche pulse generator, providing positive-going pulses to drive a laser diode. The peak current will be determined by the effective resistance of the

18   Analog circuits cookbook

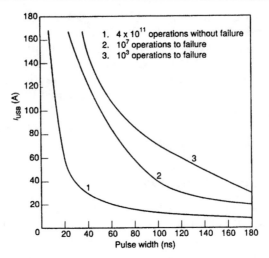

Figure 1.10 *Maximum permitted avalanche current versus pulse width for the ZTX415, for the specified reliability*

transistor in avalanche breakdown plus the slope resistance of the diode. As these two parameters are both themselves dependent upon the current, it is not easy to determine accurately just what the peak value of current is. However, this is not in practice an insuperable difficulty, for the energy dissipated in the transistor and diode is simply equal to the energy stored in the capacitor. Since, given the value of the capacitor and the supply voltage, the stored charge is known, the pulse width can be measured and the peak current estimated. If, in a particular circuit, the avalanche- and diode-slope resistances are unusually low, the peak current will be higher than otherwise, but the pulse width correspondingly narrower, the charge passed by the transistor being limited to that originally stored in the capacitor at the applied supply voltage.

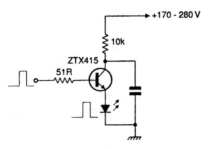

Figure 1.11 *Simple high current avalanche pulse generator circuit, driving a laser diode*

Having obtained samples of the *ZTX415*, it was decided to investigate the performance in a variant of the Figure 1.11 circuit which provides negative-going pulses, but substituting a resistive load for the diode to allow quantitative measurements to be recorded. But before commencing the tests it was necessary to find a suitable high-voltage power supply, since in these solid state days, all the ones available in the author's lab. are low-voltage types. A suitable transformer (from a long-since scrapped valve audio amplifier) was rescued just in time from a bin of surplus stock destined for the local amenity tip. It was fashioned into a high-voltage source, giving up to 800 V off-load, using modern silicon rectifier diodes. A voltmeter was included, and for versatility and unknown future applications, the transformer's low-voltage windings were also brought out to the front panel, Figure 1.12. The test set-up used is shown in Figure 1.13(a), the high-voltage supply being adjusted as required by the simple expedient of running the power supply of Figure 1.12 from a 'Regavolt' variable voltage transformer, of the type commonly known as a Variac (although the latter is a proprietary trade name).

With the low value of resistance between the base and emitter of the avalanche transistor, the breakdown voltage will be much the same as B_{VCES}, the collector-emitter breakdown voltage with the base-emitter junction short circuit. With no trigger pulses applied, the high-voltage supply was increased until pulses were produced.

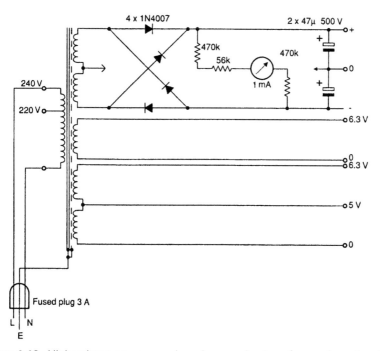

Figure 1.12 *High-voltage power supply, using a mains transformer from the days of valves*

20 Analog circuits cookbook

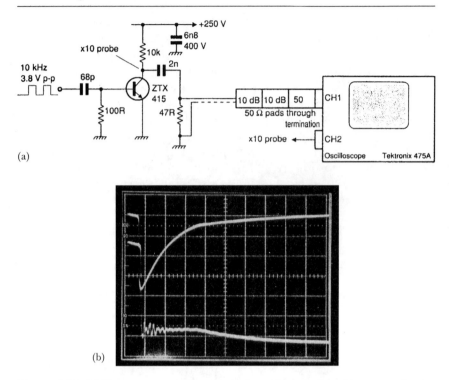

Figure 1.13 *(a) Test set-up used to view the pulse produced by an avalanche transistor. (b) Upper trace, voltage across load, effectively 50 V/div. (allowing for 20 dB pad), 0 V = 1 cm down from top of graticule, 50 ns/div.; lower trace, collector voltage, effectively 50 V/div. (allowing for ×10 probe), 0 V = 1 cm up from bottom, 50 ns/div.*

With the applied high voltage barely in excess of B_{VCES}, the prf (pulse repetition frequency) was low and the period erratic, as was to be expected. With the voltage raised further, the prf increased, the free-running rate being determined by the time constant of the collector resistor and the 2 nF capacitor. This free-running mode of operation is not generally useful, there being always a certain amount of jitter on the pulses due to the statistical nature of the exact voltage at which breakdown occurs. The high-voltage supply was therefore reduced to the point where the circuit did not free run, and a 10 kHz squarewave trigger waveform applied.

The pulses were now initiated by the positive edges of the squarewave, differentiated by the 68 pF capacitor and the base resistor, at a prf of 10 kp/s. On firing, the collector voltage drops to near zero, causing a negative-going pulse to appear across the load resistor, which consisted of a 47 Ω resistor in parallel with a 50 Ω load. The latter consisted of two 10 dB pads in series with a 50 Ω 'through termination' RS type 456-150, mounted at the oscilloscope's Channel 1 input socket and connected to the test circuit by half a metre of low loss 50 Ω coax. The cable thus presented a further 50 Ω resistive load in parallel with the 47 Ω resistor.

The drop in collector voltage can be seen to be almost the full 250 V of the supply, Figure 1.13(b), lower trace. However, the peak voltage across the load resistor (upper trace) is only around −180 V, this circuit providing a negative-going output, unlike that of Figure 1.11. The lower amplitude of the output pulse was ascribed to the ESR (equivalent series resistance) of the 2 nF capacitor, a foil type, not specifically designed for pulse operation. This is confirmed by the shape of the pulse, the decay of which is slower than would be expected from the 50 ns time constant of the capacitor and the 25 Ω load (plus transistor slope resistance in avalanche breakdown), and emphasises the care needed in component selection when designing fast laser diode circuits.

The peak pulse voltage across load corresponds to a peak current of 7.25 A and a peak power of 1.3 kW. However, the energy per pulse is only $1/2 CV^2$, where C = 2 nF and V = 250 V, namely some 63 μJ, including the losses in the capacitor's ESR and in the transistor. This represents a mean power of 630 mW, most of which will be equally divided between the 47 Ω resistor and the first of the two 10 dB pads, which is why the prf was restricted to a modest 10 kHz. The lower trace in Figure 1.13(b) shows the drop across the transistor during the pulse to be about 16 V, giving an effective device resistance in the avalanche mode of 16/7.25 or about 2.2 Ω. Thus, given a more suitable choice of 2 nF capacitor, over 90% of the available pulse energy would be delivered to the load. In the circuit of Figure 1.11, though, the laser diode slope resistance would probably be less than 25 Ω, resulting in a higher peak current, and an increased fraction of the energy lost in the transistor.

The ringing on the lower (collector) trace in Figure 1.13(b) is due to the ground lead of the ×10 probe; it could be almost entirely avoided by more careful grounding of the probe head to the circuit. As it also caused some ringing on the upper (output pulse) trace, the probe was disconnected when the upper trace was recorded, Figure 1.13(b) being a double exposure with the two traces recorded separately. The negative underswing of the collector voltage, starting 200 ns after the start of the pulse, before the collector voltage starts to recharge towards +250 V, is probably due to the negative-going trailing edge of the differentiated positive 'pip' used to trigger the transistor.

The shape of the output pulse from circuits such as Figure 1.11 and Figure 1.13(a), a step function followed immediately by an exponential display, is not ideal: for many applications, a square pulse would be preferred. This is simply arranged by using an open-circuit delay line, in place of a capacitor, as the energy storage element. When the avalanche transistor fires, its collector sees a generator with an internal impedance equal to the characteristic impedance of the line. Energy starts to be drawn from the line, which becomes empty after a period equal to twice the signal propagation time along the length of the line, as described in Ref. 4. Figure 1.14 shows three such circuits, (a) and (c) producing negative-going pulses and (b) positive going. If a long length of line is used, to produce a wide pulse, then version (b) is preferable to (a), since it has the output of the coaxial cable earthed. In (a), the pulse appears on the outer of the cable, so the capacitance to ground of the outer (which could be considerable) appears across the load. If a wide negative-going pulse is desired, then an artificial line using lumped components as in (c) can be used; here, the lumped delay line can be kept compact, keeping its capacitance to ground low. Where exceptional pulse power is required, ZTX415 avalanche transistors can be used in series to provide higher pulse voltages as in Figure 1.15(a) and (b), or in parallel to provide higher pulse currents as in (c).

A high-speed version

The risetime of the negative-going edge of the output pulse in Figure 1.13(b) was measured as 3.5 ns, or 3.2 ns, corrected for the 1.4 ns risetime of the oscilloscope. This is a speed of operation that might not have been expected from a transistor with an f_t of 40 MHz (min.) and a C_{ob} of 8 pF (max.), but this emphasises the peculiar nature of avalanche operation of a transistor. An obvious question was, could a substantially faster pulse be obtained with a higher frequency device? Low-power switching transistors, being no longer common in these days of logic ICs, the obvious

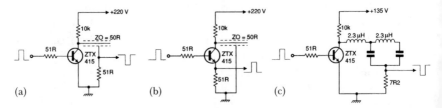

Figure 1.14 *Circuits producing square output pulses; (a) negative-going output pulses and (b) positive-going pulses both using coaxial lines; (c) negative-going pulses using a lumped component delay line*

Advanced circuit techniques, components and concepts

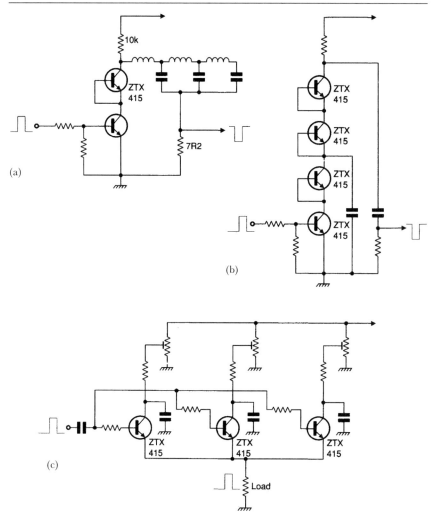

Figure 1.15 (a) A circuit for providing higher output voltage pulses. (b) A circuit for providing even higher output voltage pulses. (c) A circuit for providing higher output current pulses

alternative is an RF transistor, which will have a high f_t and a low value of C_{ob}. It was therefore decided to experiment with a *BFR91*, a device with a V_{CEO} rating of 12 V and an f_t of 5 GHz. The circuit of Figure 1.16(a) was therefore constructed, using a length of miniature 50 Ω coax, cut at random from a large reel, it turned out to be 97 cm. Given that the propagation velocity in the cable is about two thirds the speed of light, the cable represents a delay of 4.85 ns and so should provide a pulse of twice this length or, in

24 *Analog circuits cookbook*

round figures, 10 ns. Figure 1.16(b) shows (upper trace, 10 ns/div., 2 V/div., centreline = 0 V) that the circuit produced a pulse of width 10 ns and amplitude 5 V peak, into a 25 Ω load, delivering some 200 mA current. The lower trace shows (again using a double exposure) the collector voltage at 20 μs/div., 10 V/div., 0 V = bottom of graticule. With the circuit values shown, at the 20 kHz prf rate used, the line voltage has time to recharge virtually right up to the 35 V supply.

The experiment was repeated, this time with the circuit of Figure 1.17(a), the line length being reduced to 22 cm, some other component values changed and the prf raised to 100 kHz. The output pulse is shown in (b), at 1 ns/div. horizontal and >1 V/div. vertical, the VARiable Y sensitivity control being brought into play to permit the measurement of the 10% to 90% risetime. This is indicated as 1.5 ns, but the maker's risetime specification for a Tektronix 475 A

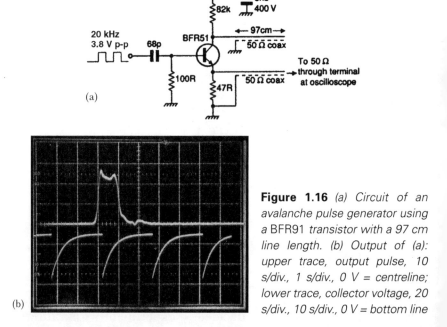

Figure 1.16 (a) Circuit of an avalanche pulse generator using a BFR91 transistor with a 97 cm line length. (b) Output of (a): upper trace, output pulse, 10 s/div., 1 s/div., 0 V = centreline; lower trace, collector voltage, 20 s/div., 10 s/div., 0 V = bottom line

oscilloscope, estimated from the 3 dB bandwidth, is 1.4 ns. Risetimes add rms-wise, so if one were to accept these figures as gospel, it would imply an actual pulse risetime of a little over 500 ps. In fact, the margin for error when an experimental result depends upon the difference of two nearly equal quantities is well known to be large.

Advanced circuit techniques, components and concepts

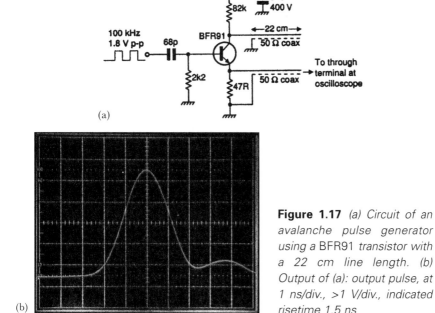

Figure 1.17 (a) Circuit of an avalanche pulse generator using a BFR91 transistor with a 22 cm line length. (b) Output of (a): output pulse, at 1 ns/div., >1 V/div., indicated risetime 1.5 ns

When the quantities must be differenced rms-wise rather than directly, the margin of error is even greater, so no quantitative certainty of the risetime in this case is possible, other than that it is probably well under 1 ns. Unfortunately, a sampling oscilloscope does not feature among my collection of test gear.

This raises the intriguing possibility that this simple pulse generator might be suitable as the sample pulse generator in a sampling add-on for any ordinary oscilloscope, extending its bandwidth (for repetitive signals) to several hundred MHz or even 1 GHz. For this application, it is important that the sample pulse generator can be successfully run over a range of repetition frequencies. With an exponential approach to the supply voltage at the firing instant, there is the possibility of jitter being introduced onto its timing, due to just how close to the supply voltage the collector has had time to recharge, see Figure 1.16(b), lower trace. The way round this is to use a lower value of collector resistance returned to a higher supply voltage. This ensures a rapid recharge, but the midpoint of the resistor is taken to a catching diode returned to the appropriate voltage just below the breakdown voltage. The collector voltage is thus clamped at a constant voltage prior to triggering, whatever the repetition rate.

References

1. Vandre, R.H. (1977) An ultrafast avalanche transistor pulser circuit. *Electronic Engineering*, mid October, p. 19.
2. NPN Silicon Planar Avalanche Transistor ZTX413 Provisional data sheet Issue 2 – March 1994.
   NPN Silicon Planar Avalanche Transistor ZTX415 Data sheet Issue 4 – November 1995.
3. The ZTX415 Avalanche Transistor Zetex plc, April 1994.
4. Hickman, I. (1993) RF reflections. *EW+WW*, October, pp. 872–876.

Negative resistance filters

Filters based on frequency-dependent negative resistors offer the performance of *LC* filters but without the bulk, expense, and component intolerance.

Filters using frequency-dependent negative resistance

Introduction

When it comes to filters, it's definitely a case of horses for courses. At RF the choices are limited; for tunable filters covering a substantial percentage bandwidth, it has to be an *LC* filter. If the tuneable elements are inductors, you have a permeability tuner; alternatively tuning may use a (ganged) variable capacitor(s), or varactor(s). Fixed frequency filters may use LCs, quartz crystals, ceramic resonators or surface acoustic wave (SAW) devices, whilst at microwaves, the 'plumbers' have all sorts of ingenious arrangements.

At audio frequencies, *LC* filters are a possibility, but the large values of inductance necessary are an embarrassment, having a poor Q and temperature coefficient, apart from their size and expense. One approach is to use '*LC*' circuits where the 'inductors' are active circuits which simulate inductance, of which there are a number, e.g. Figure 1.18. For highpass filters, synthetic inductors with one end grounded (Figure 1.18(a)) suffice, but for lowpass applications, rather more complicated circuits (Figure 1.18(b)) simulating floating inductors are required.

More recently switched capacitor filters have become available, offering a variety of filter types, such as Butterworth, Bessel, Elliptic in varying degrees of complexity up to eight or more poles. For

Advanced circuit techniques, components and concepts 27

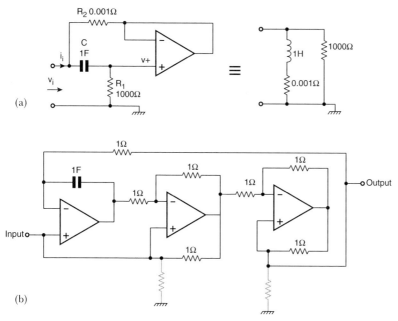

Figure 1.18 *Synthetic inductors. (a) Showing a 1 henry 'inductor' with one end grounded. Q is 10 at 0.00159 Hz, and proportional to frequency above this. Below, it tends to a 0.001 Ω resistor, just like the corresponding real inductor. (b) Floating synthetic 1 henry inductor. The high value resistors shown dotted are necessary to define the opamp dc conditions if there is no dc path to ground via Input and Output*

narrow bandpass applications, a strong contender must be the N-path filter, which uses switched capacitors but is not to be confused with switched capacitor filters; it works in an entirely different way. However, both switched capacitor and N-path filters are time-discrete circuits, with their cut-off frequency determined by a clock frequency. Hence both types need to be preceded by an anti-alias filter (and usually followed by a lowpass filter to suppress clock frequency hash). That's the downside; the upside is that tuning is easy, just change the clock frequency. The cut-off or centre frequency of a switched capacitor filter scales with clock frequency, but the bandwidth of an N-path filter does not.

Where a time continuous filter is mandatory, various topologies are available, such a Sallen and Key, Rausch, etc. An interesting and useful alternative to these and to *LC* filters (with either real or simulated inductors) is the *FDNR* filter, which makes use of frequency-dependent negative resistances.

What is an FDNR?

A negative resistance is one where, when you take one terminal positive to the other, instead of sinking current, it sources it – pushes current back out at you. As the current flows in the opposite direction to usual, Ohm's law is satisfied if you write $I = E/-R$, indicating a negative current in response to a positive pd (potential difference). This would describe a fixed (frequency-*independent*) negative resistance, but FDNRs have a further peculiarity – their resistance, reactance or impedance, call it what you will, varies with frequency. Just how is illustrated in Figure 1.19. Now with inductors (where the voltage leads the current by 90°) and capacitors (where it lags by 90°), together with resistive terminations (where the voltage leads/lags the current by 0°) you can make filters – highpass, bandpass, lowpass, whatever you want. It was pointed out in a famous paper (Ref. 1), that by substituting for L, R (termination) and C in a filter, components with 90° more phase shift and 6 dB/octave faster roll than these, exactly the same transfer function could be achieved. Referring to Figure 1.19, L, R and C are replaced on a one-for-one basis by R, C

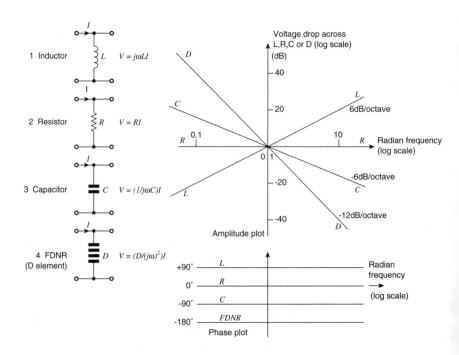

Figure 1.19 *Showing how the resistance (reactance?) of an FDNR (also known as a 'supercapacitor' or a 'D element') varies with frequency*

Advanced circuit techniques, components and concepts

and *FDNR* respectively. An *FDNR* can be realised with resistors, capacitors and opamps, as shown in Figure 1.20.

So how does an FDNR work?

Analysing the circuit of Figure 1.20 provides the answer. Looking in at node 5, one sees a negative resistance, but what is its value? First of all, note that the circuit is dc stable, because at 0 Hz (where you can forget the capacitors), A_2 has 100% NFB via R_3, and its NI (non-inverting) input is referenced to ground. Likewise, A_1 has its NI input referenced to ground (assuming there is a ground return path via node 5), and 100% NFB (A_2 is included within this loop). The clearest and easiest way to work out the ac conditions is with a vector

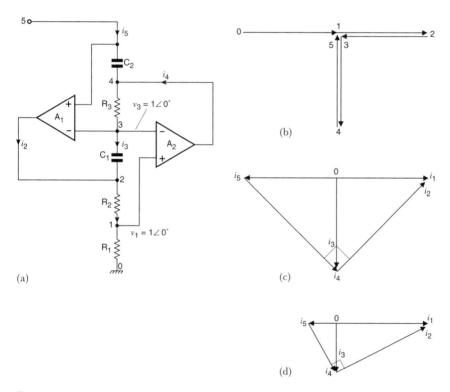

Figure 1.20 (a) FDNR circuit diagram. If v_1 is the voltage at node 1, etc., then $v_1 = v_3 = v_5$. Also, $i_1 = i_2 = i_3$ and $i_3 = i_4 + i_5$. (b) Voltage vector diagram for (a) when $R_1 = R_2 = R_3 = R$, $C_1 = C_2 = C$ and $f = \frac{1}{2\pi CR}$. (c) Current vector diagram for (a), for the same conditions as (b). (d) As (c) but for $f = \frac{1}{4\pi CR}$. Note that i_2 and i_4 are always in quadrature.

diagram; just assume a voltage at node 1 and work back to the beginning. Thus in Figure 1.20, assume that $V_{1,0}$ (the voltage at node 1 with respect to node 0 or ground) is 1 Vac, at a frequency of 1 radian per second ($1/(2\pi)$ or 0.159 Hz), and that $R_1 = R_2 = R_3 = 1\ \Omega$, $C_1 = C_2 = 1$ F. Thus the voltage at node 1 is represented in Figure 1.20(b) by the line from 0 to 1, of unit length, the corresponding current of 1 A being shown as i_1 in Figure 1.20(c).

Straight away, you can mark in, in (b), the voltage $V_{2,1}$, because $R_1 = R_2$, and node 1 is connected only to an (ideal) opamp which draws no input current. So $V_{2,1}$ equals $V_{1,0}$ as shown. But assuming A_2 is not saturated, with its output voltage stuck hard at one or other supply rail, its two input terminals must be at virtually the same voltage. So now $V_{3,2}$ can be marked in, taking one back to the same point as node 1. Given $V_{3,2}$, the voltage across C_1 (whose reactance at 0.159 Hz is 1 Ω), the current through it can be marked in as i_3 in Figure 1.20(c). Of course, the current through a capacitor leads the voltage across it, and i_3 is accordingly shown leading the voltage $V_{3,2}$ by 90°. Since $i_1 = i_2 + i_3$, i_2 can now be marked in as shown. As i_3 flows through R_3, $V_{4,3}$ can now be marked in, and as the voltages at nodes 5 and 3 must be equal, $V_{5,4}$ can also be marked in. The current i_5 through C_2 (reactance of 1 Ω) will be 1 A, leading $V_{5,4}$ as shown. Finally, as $i_3 = i_4 + i_5$, i_4 can be marked in, and the voltage and current vector diagrams (for a frequency of $1/2\pi CR$) are complete.

The diagrams show that $V_{5,0}$ is 1 V, the same as $V_{1,0}$, but i_5 flows in the *opposite* direction to i_1; the wrong way for a positive resistance. Figure 1.20(d) shows what happens at $f = 1/4\pi CR$, half the previous frequency. Because the reactance of C_1 is now 2 Ω, i_3 is only half an amp, and therefore $V_{4,3}$ is only 0.5 V. Now, there is only ½ V ($V_{5,4}$) across C_2, but its reactance has also doubled. Therefore i_5 is now only 0.25 A; not only is the current negative (a 180° phase shift), it is inversely proportional to the frequency *squared*, as shown for the FDNR in Figure 1.19.

Pinning down the numbers

Looking in at node 5, then, appears like a $-1\ \Omega$ resistor at 0.159 Hz, but you need to know how this ties up with the component values. The values of the vectors can be marked in, on Figure 1.20(b) and (c), starting with $V_{1,0} = 1$ V. Then $V_{2,1} = R_2/R_1$, and $V_{3,2} = -R_2/R_1$. It follows that $i_3 = (-R_2/R_1)/(1/j\omega C_1) = -j\omega C_1 \cdot R_2/R_1$. $V4,3 = R_3\ i_3 = -j\omega C_1 \cdot R_2 \cdot R_3/R_1$, and $V5,4 = -V4,3$. So $i_5 = -V4,3/(1/j\omega C_2) = j\omega C_1 j\omega C_2 \cdot R_2 \cdot R_3/R_1$. Looking in at node 5 the resistance is $V_{5,0}/i_5 = V_{1,0}/i_5$, where $V_{1,0} = 1$ V. So finally the FDNR input looks like:

$$FDNR = R_1/(j\omega C_1 j\omega C_2 \cdot R_2 \cdot R_3) = -R_1/(\omega^2 \cdot C_1 \cdot C_2 \cdot R_2 \cdot R_3) \qquad (1.1)$$

With 1 Ω resistors and 1 F capacitors, this comes to just -1 Ω at $\omega =$ 1 radian per second or 0.159 Hz. To get a different value of negative resistance at that frequency, clearly any of the Rs or Cs could be changed to do the job, but it is best to keep all the Rs (at least roughly) equal, and the same goes for the Cs. As a cross-check on equation (1.1), note that it is dimensionally correct. The units of a time constant CR are seconds, whilst the units of frequency are 1 per second (be it cycles or radians per second). Thus the units in the denominator cancel out, and with a dimensionless denominator, the expression has the units of the numerator R_1, ohms.

A practical example

Designing an *FDNR* filter starts off with choosing an *LC* prototype. Let's consider a simple example; a lowpass filter with the minimum number of components, which must reach an attenuation of 36 dB at little more than twice the cut-off frequency. This is a fairly tall order, but a three pole elliptic filter will do the job, if we allow as much as 1 dB passband ripple. A little experimentation with a CAD program came up with the design in Figure 1.21(a). Nice round component values, although the cut-off frequency is just a fraction below the design aim of 1 radian per second.

If you were designing an *LC* filter as such, you would certainly choose the π section of Figure 1.21(b) rather than the TEE-section, as the π section is the minimum inductor version. But for an *FDNR* filter, the minimum capacitor version is preferable, as the Cs become *FDNR*s (fairly complicated), whereas the Ls become Rs and are therefore cheap and easy. But before passing on to consider the *FDNR*, note that the computed frequency response of the normalised 1 Ω impedance *LC* filter is as shown in Figure 1.21(c). The low-frequency attenuation shows as 6 dB rather than 0 dB, because the 1 Ω impedance of the matched source (a 2 V emf ideal generator behind 1 Ω) is considered here as part of the filter, not as part of the source. To the 2 V generator emf (which is what the CAD program models as the input), the source and load impedance appear as a 6 dB potential divider.

The *FDNR* version of the filter is shown in Figure 1.22 – not only do the Ls become Rs and the Cs *FDNR*s, but the source and termination resistors become capacitors. In an *LC* filter, the source and terminating resistors would usually be actually part of the source and load respectively. But an *FDNR* filter at audio frequencies will be driven from the 'zero' output impedance of an opamp and feed into

32 Analog circuits cookbook

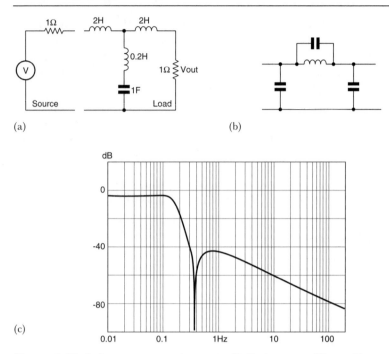

Figure 1.21 *A low component count elliptic lowpass filter with a minimum attenuation of 36 dB from twice the cut-off frequency upwards, the price being as much as 1 dB passband ripple. The minimum capacitor design of (a) is more convenient than (b) for conversion to an FDNR filter. (c) The frequency response of the filter*

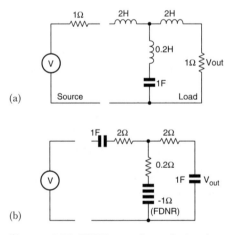

Figure 1.22 *FDNR version of the lowpass filter*

the near infinite impedance of another, so you must provide the terminations separately if you want the response to be the same as the prototype LC filter. In Figure 1.22, the inductors have been replaced with resistors on an ohm per henry basis, and the Rs and Cs converted to Cs and $FDNR$s similarly. As it happens, the required $FDNR$ value is -1 Ω, so values of 1 Ω and 1 F in the circuit of Figure 1.20 will do the job. Had one used the tabulated values for a 1 dB ripple 35 dB A_s three pole filter, e.g. from Ref. 2 (see Figure 1.23), the required value of C_2 in the TEE-section version would have been 0.865 F. Accordingly, from equation (1.1), R_1 in Figure 1.20 would become 0.865 Ω, or you could alternatively change R_2 and/or R_3 to achieve the same effect. Or you could scale C_1 and/or C_2 instead, but it is best to leave them at 1 F – the reason for this will become clear later.

Having arrived at a 'normalised' FDNR filter design (i.e. one with a 0.159 Hz cut-off frequency), the next step is to denormalise it to the wanted cut-off frequency, let's say 10 kHz in this case. No need to change the Rs at this stage, but to make the FDNR look like −1 Ω (or −0.865 or whatever) at 10 kHz, the capacitor values must be divided by 2π times ten thousand. And since the termination capacitors must also look like 1 Ω at this frequency, they must be scaled by the same ratio. You now have a filter with the desired response and cut-off frequency, but the component values (shown in round brackets in Figure 1.24) are a little impractical. This is easily fixed by a further stage of scaling. Since resistors are more easily obtainable in E96 values and 1% selection tolerance, it pays to scale the 15.9 µF capacitors to a nice round value – say 10 nF. So all impedances must be increased by this same ratio $N = 1590$; the resistors multiplied by N and the capacitors

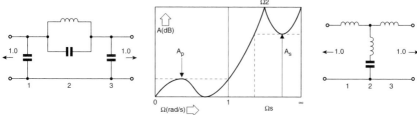

$A_p = 1$ db

Ω_s	A_s [db]	C_1	C_2	L_2	Ω_2	C_3
1.295	20	1.570	0.805	0.613	1.424	1.570
1.484	25	1.688	0.497	0.729	1.660	1.688
1.732	30	1.783	0.322	0.812	1.954	1.783
2.048	35	1.852	0.214	0.865	2.324	1.852
2.418	40	1.910	0.145	0.905	2.762	1.910
2.856	45	1.965	0.101	0.929	3.279	1.965
Ω_s	A_s [db]	L_1	L_2	C_2	Ω_2	L_3

(© 1958 IRE (now IEEE))

Figure 1.23 Tabulated normalised component values for three pole 1 dB passband ripple elliptic filters with various values of As at Ω_s (here Ω means the same as ω elsewhere in the article)

34 Analog circuits cookbook

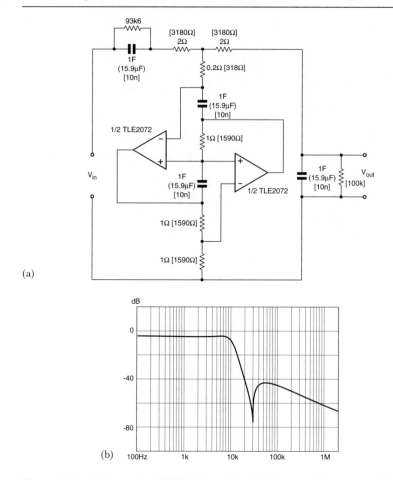

Figure 1.24 (a) Complete FDNR filter with 10 kHz cut-off frequency. The (values) are a little impractical, but are easily scaled to more sensible [values]. (b) Computed frequency response of the above filter. The cut-off frequency (at −1 dB) is a shade below the intended value, as was that in Figure 1.21(c)

divided by N. Conveniently, the Cs in the $FDNR$ are the same value as the terminating capacitors, if (as recommended) any change in the required normalised $FDNR$ negative resistance was effected by changing the R values only. The resultant practical component values are shown in square brackets in Figure 1.24.

One peculiarity of an $FDNR$ filter is due to its use of capacitive terminations. The impedance of these varies with frequency and, notably, becomes infinite at dc (0 Hz). Thus any practical $FDNR$ filter would have infinite insertion loss at this frequency! This is remedied by connecting resistors in parallel with the terminating capacitors, to determine the 0 Hz response. They are shown in Figure 1.24(a) and have been chosen (taking into account the two 3180 Ω resistors) to provide 6 dB attenuation at 0 Hz, to match the filter's passband 6 dB loss. With the addition of these, Figure 1.24(a) is now a practical, fully paid-up working lowpass filter, the computed frequency response of which is shown in Figure 1.24(b). This is all fine in theory, but does it work in practice?

Proof of the pudding

Ever of a pragmatic (not to say sceptical) turn of mind, I determined to try it out for real. So the circuit of Figure 1.24(a) was made up (almost) exactly as shown, and tested using an HP3580A audio frequency spectrum analyser, the circuit being driven from the *3580A*'s internal tracking generator. There were minor differences; whereas the plot of Figure 1.24(b) was modelled with *LM318* opamps, these were not to hand. So a *TLC2072CP* low-noise, high-speed J-FET input dual opamp was used, a very handy Texas Instruments device with a 35 V/μs slew rate and accepting supplies in the range ±2.25 V to ±19 V. The required resistor values were made up using combinations of preferred values, e.g. 82K + 12K for 93.6K, 270R + 47R for 318R, etc., all nominal values thus obtained being within better than 1% of the exact values. 100K + 12K was used for the terminating resistor, to allow for the 1M input resistance of the spectrum analyser in parallel with it. The resistors were a mixture of 1% and 2% metal film types, except the 47R which was 5%. The four 10 nF capacitors were all 2.5% tolerance polystyrene types.

Having constructed the circuit and powered it up, it didn't work. A quick check with a 'scope showed that the opamp output pin was stuck at +10 V, with various other pins at peculiar voltages. The connections were all carefully checked and found correct, leaving little room to doubt that the opamp was at fault. But this always rings alarm bells with me, as in 99.9% of cases, when a circuit sulks it is not the fault of a component, but a blunder on the part of the constructor. Still, the opamp was removed and another sought. At this point I realised that the offending item was in fact a *TLE2027* (a single opamp), not a *TLE2072*, and remembered with a mental grimace that this was the second time I had fallen into this elementary trap.

With the right opamp in place, the circuit worked, but the response was not exactly as hoped, due to being driven from the *3580A*'s 600 Ω source impedance. So the *TLE2027* (which had survived its misconnected ordeal unscathed) was redeployed as a unity gain buffer to drive the filter from a near zero source impedance, and its output level set at top-of-screen. The results are shown in Figure

36 *Analog circuits cookbook*

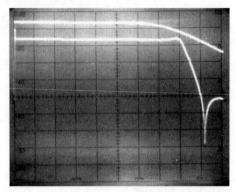

Figure 1.25 *Actual frequency response of the circuit of Figure 1.24. The horizontal scale is log frequency, the left-hand vertical being 20 Hz, the 3rd, 6th and 9th vertical graticule lines representing 200 Hz, 2 kHz and 20 kHz respectively. Horizontal graticule lines are at 10 dB intervals. Upper trace, generator reference level top of screen, representing the source emf. This trace was recorded with the shunt leg of the filter open circuited (318 Ω resistor removed). Lower trace, response of complete filter (318 Ω resistor replaced). Reference level has been moved down one graticule division for clarity.*

1.25. First, the filter action was disabled by removing the 318 Ω resistor, leaving a straight-through signal path. The upper trace shows the 6 dB loss due to the terminations, mentioned earlier, and also a first-order roll-off due to the effect of the terminating capacitor at the load end, with the two 318 Ω resistors. The lower trace shows the response of the complete filter (318 Ω resistor replaced). The reference level has been moved down one graticule division for clarity. The −1 dB point is at two divisions in from the right, which given the horizontal scaling of three divisions per decade, corresponds to 9.3 kHz, pretty close agreement with the predicted performance of Figure 1.24(b). In log frequency mode, the analyser's bandwidth extends only up to 44.3 kHz, but this is far enough to see that the notch frequency and the level of the return above it (36 dB below the LF response) also agree with the computed results.

Others like it, too

FDNR filters have found various applications, especially in measuring instruments. The advantage here is that the response is predictable and close to the theoretical. Some other active filter sections (e.g. Sallen and Key), when combined to synthesise higher order filters, show a higher sensitivity to component tolerances. This is a disadvantage where the filters are used in the two input channels of an instrument which requires close matching of the channel phase and amplitude responses. For this reason, FDNR filters (see Figure 1.26) were used in the input sections of the *HP5420A* (Ref. 3).

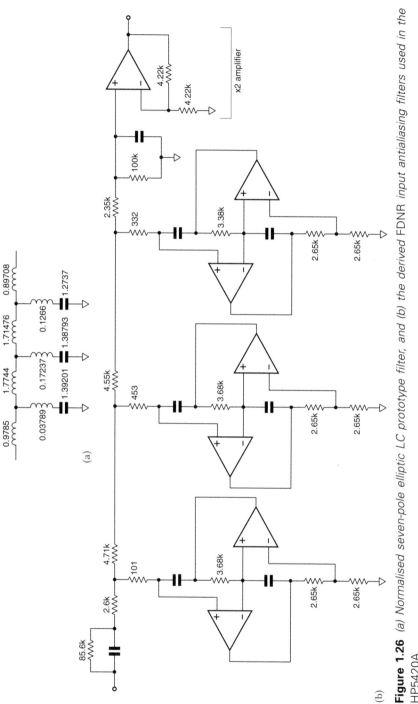

Figure 1.26 (a) Normalised seven-pole elliptic LC prototype filter, and (b) the derived FDNR input antialiasing filters used in the HP5420A

Log sweeps and IF bandwidths

The response shown in Figure 1.25 was taken using the log frequency base mode of the *HP3580A* 0–50 kHz spectrum analyser. In this mode, the spot writes the trace across the screen at a steady rate, taking about 6 seconds to sweep from 20 Hz to 44.3 kHz. Thus the sweep rate in Hz per second increases greatly as the spot progresses across the screen. This means that if a resolution bandwidth narrow enough to resolve frequency components encountered near the start of the sweep (e.g. 1 or 3 Hz bandwidth) is used, near the end of the sweep the analyser will be passing through any signals far too fast to record their level even approximately. On the other hand, if a bandwidth wide enough to accurately record signal amplitudes in the 20 kHz region (such as 300 Hz) is used, the zero frequency carrier breakthrough response will extend half way across the screen. So, when using log sweep mode to record the amplitudes of stationary signals, compromises must be made.

But this is not the case in Figure 1.25. For here, the only signal of interest is the output of the tracking generator, to which the analyser is, by definition, always tuned. So the analyser is at no time sweeping through a signal and in principle it might seem that the 1 Hz bandwidth could be used. There is a restraint on the bandwidth, however, set by the rate at which the signal amplitude changes. This can get quite fast in the vicinity of a notch, and accordingly the trace in Figure 1.25 was recorded with a 30 Hz resolution bandwidth. At 10 Hz bandwidth, the notch appeared shunted slightly to the right and its full depth was not recorded. On the other hand, at a 100 Hz bandwidth the notch response was identical to that shown, but the left-hand end of the trace, representing 20 Hz, was elevated slightly, due to the zero frequency carrier breakthrough response. If, due to a fortuitous conjunction of component tolerances, the actual notch depth had been much deeper than it actually was, the 100 Hz bandwidth would have been necessary to capture it. In that case, it would be better to switch back to linear frequency base mode, and make the notch measurement at a span of 100 Hz – or even 10 Hz – per horizontal division.

References

1. Bruton, L.T. (1969) Network transfer functions using the concept of frequency dependent negative resistance. *IEEE Transactions on Circuit Theory*, Vol. CT-16, August, pp. 405–408.

2. Hickman, I. (1993) *Newnes Practical RF Handbook*. ISBN 0 7506 0871 4, p. 245.
3. Patkay, Chu and Wiggers (1977) Front-end design for digital signal analysis. *Hewlett Packard Journal*, October, Vol. 29, No. 22, p. 9.

Tiny components

At one end of the surface-mount spectrum, complex digital ICs are becoming so densely pinned that they make prototyping almost impossible. At the other, it is now easy to obtain one logic function or opamp in a single, minute sm package. While reducing product size, these tiny devices can simplify implementation, improve performance, and even open up new application areas.

Big surprises ... in small packages

Introduction

The surface-mount revolution has been under way for years now, with most products using surface-mount passives. Fixed resistors are migrating from the 1208 size (0.12 by 0.08 inches) to 0805, 0604 or even 0402. Trimmer resistors, with overall dimensions of less than 4 mm^2, are supplied by several manufacturers, including Bourns and Citec. Capacitors are available in a similar range of sizes to fixed resistors, though the larger values such as tantalum electrolytics tend to be in 1208 format still, or larger, for obvious reasons. Trimmer capacitors are available with a footprint of less than 4 mm^2, from various manufacturers, e.g. Murata. Surface-mount inductors are available in the various formats, whilst ingenious surface-mount carriers accommodate ferrite toroid cored inductors where higher values of current-carrying capacity or of inductance are necessary – such as in switchmode power supplies – and where the extra height can be accommodated.

But surface-mount passives have been around so long that there is not much new to say about them. So this article concentrates on active devices, and mainly on integrated circuits, ICs, in particular, which is where the action currently is. In the following, various aspects of the application of these devices is discussed, and just a few of the many hundreds of types available are briefly presented.

The packages

For years, ICs came in just two widths, and a variety of lengths, all with pins on 0.1in. centres. Thus 8, 14 and 16 pin dual-in-line (DIL) devices (whether side brazed ceramic types to military specifications, or commercial plastic 'DIPs') came with a width between rows of pins of 0.3in. But 0.6in. was the order of the day for ICs with 24, 28, 40 or 68 pins. Even so, there were exceptions, such as 0.3′ 'skinny' 24 pin devices. But then, with the appearance of more and more complex ICs, more and more pin-outs were necessary. To accommodate these, square devices with pins on all four sides appeared, such as chip-carriers – both leadless and leaded – J lead devices and plastic quad flatpacks (PQFP) with various pin centre spacings, often only 0.025′ or less, and up to 200 pins or more. To minimise the package size, ICs were packaged in 'pin-grid array' packages with several parallel rows of pins on the underside of each edge, and again up to 200 or more pins. Yet other formats are SIL/SIP (single-in-line/plastic) packages for memory chips and surface-mount audio frequency power amplifiers. AF PAs also appear in through-hole mounting SIPs, with alternate pins bent down at different lengths, to mount in two rows of staggered holes.

More recently, there has been renewed interest in really tiny devices with eight, five or even just three pins. This format has long been favoured by RF engineers for UHF and microwave transistors, the consequent reduction in overall size and lead lengths contributing to minimal package parasitics. Now the advantages of really tiny devices, which are many, are becoming available also to analog and digital designers, and this article looks at some of these devices. Table 1.1 lists typical examples, giving the package designation (which varies somewhat from manufacturer to manufacturer), the number of pins, a typical example of a device in that package, and its manufacturer, and the maximum overall size of the 'footprint' or board area occupied by a device in that package style (this again varies slightly from manufacturer to manufacturer).

With devices in such small packages, getting the heat away can be a problem. With many of these ICs, though, the difficulty is alleviated due to two aspects. First, many devices such as opamps, comparators and digital ICs now work from a single supply of 3 V or even lower, as against the 5 V, ±5 V or even ±15 V required by earlier generations. Second, with improved design techniques, high-speed wide frequency range devices can now be designed to use less current than formerly. Nevertheless, thermal considerations still loom large in many cases, when applying these tiny devices. This is discussed further in the following sections, which deal with various classes of small outline devices.

Advanced circuit techniques, components and concepts

Table 1.1 Some representative devices in small packages, from various manufacturers

Style	Leads	Example	Function	Manufacturer	Footprint max.
SOD-323	2	1SS356	Diode, band-switching	Rohm	1.35 x 2.7 mm
SOT23-3 ('TinyPak'™. Also known as TO-236-AB)	3	LM4040AIM3–5.0	Voltage ref. 5 V 0.1%	Nat. Semi.	3.0 x 3.05 mm
SOT23-5 (JEDEC TO-xxxxx outline definition now due)	5	AD8531ART	Opamp, 5 V, 250 mA op.	Analog Devices	3.0 x 3.1 mm
SO-8	8	MAX840	–2V regulated GaAsFET Bias generator	MAXIM	5.03 x 6.29 mm
SO-14	14	LT1491CS	Quad opamp, 2–44 V supply	LINEAR Tech	6.20 x 8.74 mm

SO = 'small outline'

Discrete active devices

With discretes such as diodes, in many cases maximum dissipation is a pressing consideration, and package styles and sizes reflect this. Thus the UDZ series zeners from Rohm, in the SOD-323 package, Figure 1.27(a), are rated at 200 mW. But RLZ series devices (also from Rohm) in the slightly larger LL34 package (Figure 1.27(b)) dissipate 500 mW, while their PTZ series in the even larger PSM package (Figure 1.27(c)) are rated at 1 W.

With active devices also, special packages are used to cope with the device dissipation. For example, the *IRFD11x* series MOSFETs are mounted in a four pin 0.3′ DIL package, see Figure 1.28(a). Pins 3 and 4 are commoned and provide not only the drain connection, but also conduct heat to through-hole pads (hopefully of generous dimensions) on the PCB, providing a P_{drain} rating of 1.2 W. This is actually 20% more than the rating of the VN10 KM, which is housed in a TO237 package, see Figure 1.28(b) – this is like a TO92, but with a metal tab, connected to the drain, projecting from the top. The SOT89 is an even smaller package (Figure 1.28(c)), measuring just 2.5 mm by 4 mm, excluding leadouts. Nevertheless, the Rohm *BCX53* is rated at 500 mW, or 1 W when mounted on a suitable ceramic PCB. The wider collector lead, on the opposite side of the package from the base and emitter leads, bends back under the body of the device, providing a large heat transfer area. The SOT223 package (not shown) provides a power dissipation of up to about 1.5 W at 25°C. The TO252 'D-pak' (Figure 1.28(d)) – housing, for example, the *IFRF024*, a 60 V 15 A MOSFET with a 60 A pulsed I_d rating – does even better. The device dissipates watts, if you can keep its case temperature down to 25°C.

42 Analog circuits cookbook

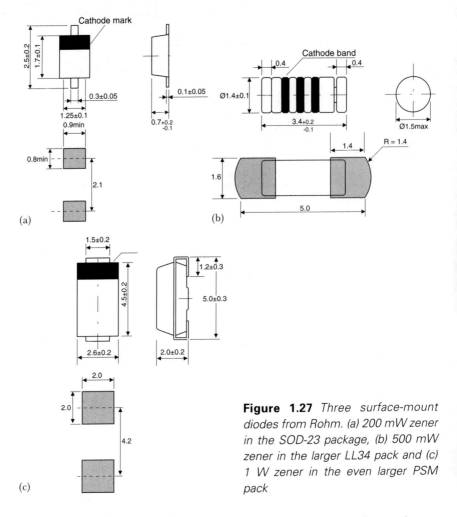

Figure 1.27 *Three surface-mount diodes from Rohm. (a) 200 mW zener in the SOD-23 package, (b) 500 mW zener in the larger LL34 pack and (c) 1 W zener in the even larger PSM pack*

For small signal amplifiers, size is less important and transistors are available in packages smaller than SOT23 (SMT3), Figure 1.29(a). The UMT3 (Ultramold, SOT323) package of Figure 1.29(b) has a footprint of 2.2 × 2.2 mm overall, including leads, whilst the EMT3 (Figure 1.29(c)) occupies just under 1.8 × 1.8 mm overall, these being the maximum dimensions. With such very small devices, traditional lab prototyping becomes very difficult, not to say tedious.

Analog ICs

With digital ICs, the trend is to higher and higher levels of functional integration, with an inevitable accompanying inflation in the number

Advanced circuit techniques, components and concepts   43

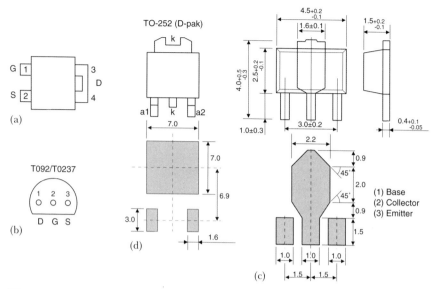

Figure 1.28 (a) Four pin 0.2' DIP package often used for FETs and other small-power devices; (b) the TO237 pack is like a TO93, but with a small metal tab extending from the top; (c) the SOT89 pack can typically dissipate 0.5–1 W; (d) the TO252 package dissipates watts – at least at 25°C case temperature!

of pins per package. In the analog world, however, general-purpose functions, such as opamp, comparator, buffer, voltage reference, etc., tend to dominate. The result is that whilst digital ICs tend to get bigger (or at least not much smaller, due to all those pins), analog functions are appearing in smaller and smaller packages. The exception is DACs and ADCs with parallel data buses. But these ICs tend to bridge the analog/digital divide anyway, and even here, devices in tiny 8 pin packages are readily available, thanks to the economy in pin numbers afforded by using serial data input/output schemes rather than bus structures.

Whilst single transistors can be mounted in packages smaller than SOT23, this is more problematical for the larger silicon die of ICs. So for the most part, the 3 pin version of SOT23 is the smallest package used for ICs. An example is the *AD1580* 1.2 V micropower precision shunt voltage reference from Analog Devices. To the user, this appears simply as a 1.2 V zener diode. But the dynamic output impedance (ac slope resistance) at 1 mA is typically just 0.4 Ω, resulting in a change in output voltage, over 50 μA to 1 mA and over −65 to +125°C, of only 500 μV typical. Being a two terminal device, pin 3 is no connection, or may be connected to V−.

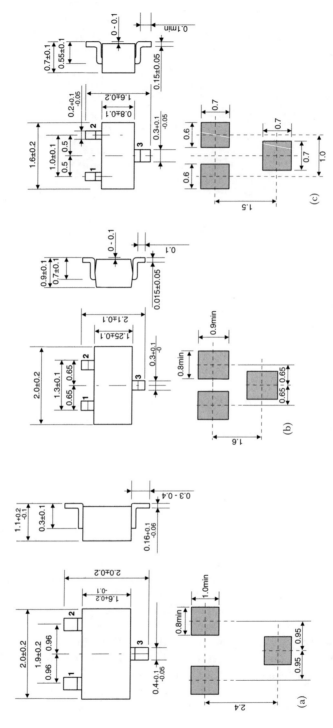

Figure 1.29 Three small transistor outlines: the tiny SOT23-3 (a) dwarfs the SOT323 (b), which in turn dwarfs the minuscule EMT3 (c)

Advanced circuit techniques, components and concepts 45

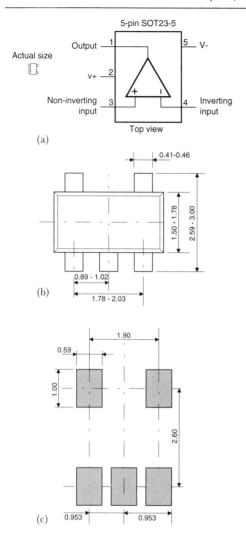

Figure 1.30 The LMC7111 from National Semiconductor. (a) Pinout and actual size; (b) dimensions of the SOT23-5 package, and of the recommended PC pads

A good example of an opamp in a small package (also available in an 8 pin DIP) is the *LMC7111* from National Semiconductor, Figure 1.30. The leadout arrangement of the 5 pin SOT23-5 version is shown in Figure 1.30(a): note the actual size drawing alongside! The device is a CMOS opamp with rail-to-rail input and output, operating from a supply voltage V_s of 2.7 V upwards (absolute max. 11 V). With a gain/bandwidth (GBW) product of 40 kHz with a 2.7 V supply, it draws a supply current I_s of around 50 μA. Its bipolar stablemate, the *LMC7101*, offers a 0.6 MHz GBW and 0.7 V/μs slew rate in exchange for an I_s of around 800 μA, also at 2.7 V.

Where something a little faster is needed, then in the same package and from the same manufacturer comes the *LM7131* high-speed bipolar opamp. This has a GBW of 70 MHz, and a slew rate of 100 V/μs, even when driving a capacitive load of 20 pF. Total harmonic distortion at 4 MHz is typically only 0.1% when driving a 150 Ω load with a 3 V V_s, and even with this level of performance, I_s is only 8 mA.

Where blindingly fast speed is necessary, the *LM7121* voltage feedback opamp, in the same package with the same pinout, has a 1300 V/μs slew rate, for an I_s of just over 5 mA typical. But note that this is the performance with dual supplies of +15 and −15 V. The device works on a single V_s of down to 5 V, but the performance is then more modest. Unusually for an opamp, this device is stable with literally any level of load capacitance, maximum peaking (up to 15 dB) occurring with around 10 nF. Other stablemates in the same SOT23 package and with the same pinouts are the *LMC7211* and *LMC7221* rail-to-rail input comparators, with active and open drain outputs respectively.

Current feedback opamps are known for their excellent ac characteristics. The *OPA658* is a wideband low-power current feedback opamp from Burr-Brown, available in the SOT23-5 pin package. With a unity gain stable bandwidth of 900 MHz and a 1700 V/μs slew rate, it has a wide range of applications including high-resolution video and signal processing, where its 0.1 dB gain flatness to 135 MHz is exceptional.

Where a circuit requires two opamps, two devices in, say, SOT23-5 packages may be used, and this provides the ultimate in layout flexibility. It may even take up less space than a dual, but the dual opamp will usually be cheaper than two singles. Figure 1.31 shows the *AD8532* dual rail-to-rail input and output CMOS opamp from Analog Devices. Featuring an output drive capability of a quarter of an amp and a 3 MHz GBW (at $V_s = 5$ V), it operates from a single supply in the range 2.7 to 6 V. Figure 1.31(a) and (b) compare the footprint in the TSSOP (thin shrink small outline package) and the SO-8 package. Width over the pins is similar, but the TSSOP's pin spacing of 0.65 mm, against twice this for the SO-8, results in a package length not much more than half that of the SO-8. (For applications where more space is available, the device also comes in the old-fashioned 8 pin DIP package.)

Figure 1.31(c) shows the opamp's internal circuitry (simplified). As common in devices with a rail-to-rail input, whether bipolar or FET, complementary input pairs in parallel are used. Likewise, for rail-to-rail outputs, common drain (collector) stages are dropped in favour of common source (emitter) stages. Figure 1.31(d) shows the clean large-signal pulse response, even at a V_s of just 2.7 V. The device is just one of the family of AD8531/2/4 single/dual/quad opamps, available in a wide variety of package styles.

Another dual opamp, this time with the exceptional V_s range of 2.7 to 36 V, is the *OPA2237*, from Burr-Brown. With its maximum offset voltage of 750 μV and its 1.5 MHz bandwidth, it is targeted at battery-powered instruments, PCMCIA cards, medical instruments, etc. It is available in SO-8, and also in MSOP-8 (micro small outline package) which is just half the size of the SO-8 package.

Advanced circuit techniques, components and concepts 47

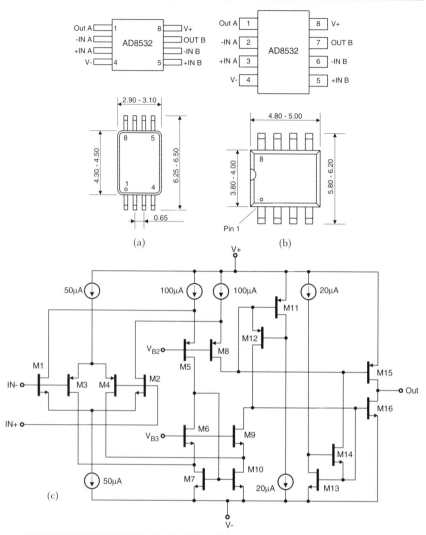

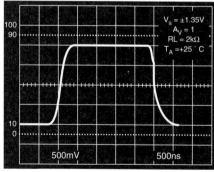

Figure 1.31 The AD8532 dual opamp from Analog Devices is available in TSSOP (a), SO-8 (b) or 8 pin DIP. The paralleled complementary input stages and common source output stages provided rail-to-rail operation at both ends (c). The 2 V peak-to-peak response, operating on ±1.35 V rails, is shown in (d)

48 Analog circuits cookbook

Other analog circuits

Figure 1.32 shows the MAXIM *MAX8865x* dual low drop-out regulator, where suffix x is T, S or R, indicating preset output voltages of 3.15, 2.84 or 2.80 V respectively. Each output is capable of supplying up to 100 mA, with its own individual shutdown input. Figure 1.32(a) shows the device connected to supply output 1 continuously, and output 2 only when the SHDN2 bar pin is high. If the SET1 (or the SET2) pin is connected not to ground, but to a voltage divider connected across the corresponding output, the circuit will produce whatever stabilised output voltage results in the SET pin being at 1.25 V (assuming, of course, that the input voltage, which must be in the range 2.5 to 5.5 V, is adequate). Internal circuitry for each output senses whether the SET pin is at a voltage below or above 60 mV, and selects an internal, or the external voltage divider respectively. The pin allocation is as in Figure 1.32(b), whilst the package dimensions are given in (c). This package is proprietary to MAXIM, being the same length as an 8 pin TSSOP, but with a narrower body, making the width over pins rather smaller. The *MAX8866* is similar, but includes an auto-discharge function, which discharges an output to ground whenever it is deselected.

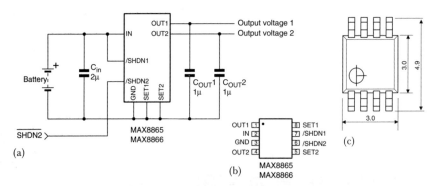

Figure 1.32 *The MAX8865x dual low dropout regulator (a) from MAXIM comes in the proprietary muMAX package, with pinout as at (b). At 3 mm, the package length (c) is similar to TSSOP, but the width across pins is 1.5 mm less, which could lead to its more widespread adoption by other manufacturers*

Figure 1.33 shows two other MAXIM devices. At (a) are shown the *MAX4051* and *MAX4052*, these being single NO and N/C analog switches respectively. Mounted in SOT23-5 packages, they are used where a single switch function is needed, providing it in much less

Advanced circuit techniques, components and concepts 49

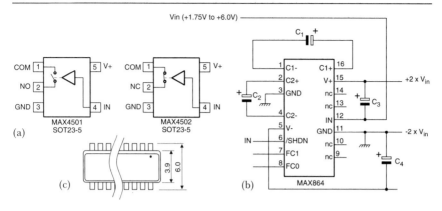

Figure 1.33 (a) Single NO or NC analog switches save space compared to leaving a quarter of a quad pack unused

space than would be occupied by a quad analog switch pack. At (b) is shown the *MAX864* dual-output charge pump. This provides outputs of $+2\,V_{in}$ and $-2\,V_{in}$ nominal, for any input V_{in} in the range $+1.75$ to $+6.0$ V. Two pins, FC0 and FC1, are connected to ground or V_{in} as required, offering a choice of four different internal switching frequencies in the range 7 to 185 kHz, assuming that the SHDN bar pin is high. The *MAX864* is packaged in a QSOP outline, see Figure 1.33(c).

Figure 1.34 shows a 12 bit DAC, the *LTC1405/1405L*, from Linear Technology. It accepts 12 bit parallel input data and the *LTC1405* outputs up to 4.095 or 2.048 V (pin strappable selection), from a 4.5 to 5.5 V supply. The *LTC1450L* provides a 12 bit resolution output of up to 2.5 or 1.22 V, from a 2.7 to 5.5 V supply. Figure 1.34(a) shows the internal workings of the chip, which is available mounted in a 24 lead SSOP package, (b), or in a 28 pin DIP. Figure 1.34(c) shows the companion *LTC1458/1458L*, which is a quad 12 bit DAC. It is shoehorned into a 28 pin SO package, or a 28 pin SSOP, by using a serial data input scheme, rather than the parallel data input of the *LTC1405/1405L*.

Figure 1.35 shows another DAC, this time one which accepts 16 or 18 bit data and designed for use in CD systems, MPEG audio, MIDI applications, etc. The *PCM1717E* from Burr-Brown incorporates an × 8 oversampling digital filter, multilevel delta-sigma DAC and analog lowpass in each of its stereo output channels. Its selectable functions include soft mute, digital de-emphasis and 256 step digital attenuation. Using a serial data input, it is supplied in a 20 pin SSOP package, a shorter version of that shown in Figure 1.34(b).

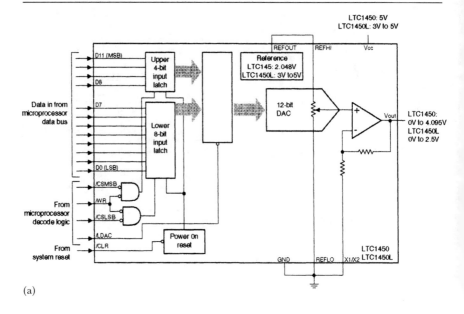

(a)

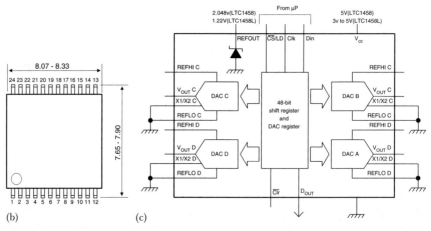

(b)   (c)

Figure 1.34 *The LTC1405, (a), from Linear Technology is a 12 bit DAC with parallel data input. This requires a 24 pin package, (b), but the Small Outline pack is still much smaller than the corresponding DIP. (c) shows a block diagram of the internal workings of the LTC1458, from the same manufacturer. This quad DAC comes in either an SO pack, or the even smaller SSOP, both with 28 pins, achieved by using a 48 bit serial data input stream*

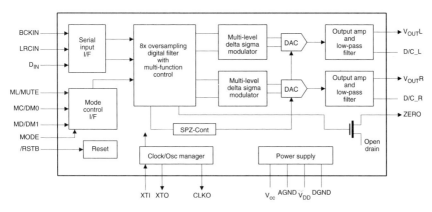

Figure 1.35 *The PCM1717E DAC from Burr-Brown accepts 16 or 18 bit serial data, and provides L and R stereo output channels. With numerous facilities, aimed at CD systems, MPEG audio, MIDI applications, etc.*

Digital circuits

Traditional small scale integration SSI, and MSI logic circuits – originally supplied in 0.3in. width packages with up to 16 (later, 18, 22 or more) pins – have long ago migrated to the SO package and even smaller packs. LSI devices with up to 64 or 68 pins came in 0.6in. wide packs, but then migrated to a variety of package types, including leaded and leadless chip carriers, J lead packs, pin grid arrays, etc., with the latest development being ball pin arrays. But processors, DSP chips and the like tend to require so many leadouts that they hardly come under the heading of tiny devices, even though truly small considering the number of pins. This is illustrated in Figure 1.36, which shows packages with a modest 44 pins, (c) and (d); 52 pins, (b); and 240 pins, (a). This latter package even comes in a version with 304 pins.

In addition to processors, DSP chips, etc., package types with a large number of pins are also used for custom- and semi-custom logic devices, and programmable arrays of various types. These enable all the logic functions associated with a product to be swept up into a single device, reducing the size and cost of products which are produced in huge quantities. But this approach is not without its drawbacks, often leading to practical difficulties at the layout stage. For example, on a 'busy' densely packed board, the odd logic function such as an inverter, AND gate or whatever, may be required at the opposite end of the board from that at which the huge do-it-all logic package is situated. This forces the designer either to accept long digital signal runs right across the board, or to include a quad SSI

52 Analog circuits cookbook

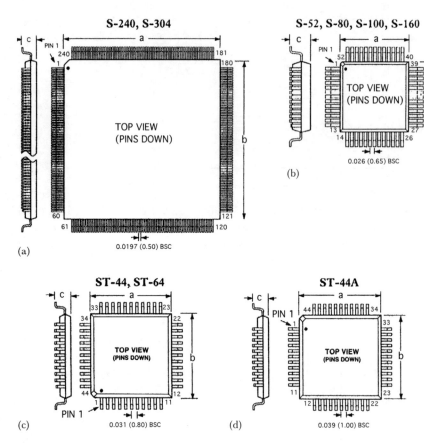

Figure 1.36 *Digital ICs come in packs with up to 300 or more pins. (a) shows the 240 pin PQFP (plastic quad flat package) S-240. The slightly wider pin spacing of PQFP packs with up to 160 pins, (b), is more manageable. There are traps for the unwary! The two 44 pin TQFP (thin quad flat package) packs in (c) and (d) look very similar, but the pin spacing is different*

package of which only a quarter is used, or to seek some other solution.

Such a solution is now at hand, right at the other extreme from multi-pin packs, or even 14 pin SSI quad gate packs. For example, a simple RTL (resistor/transistor logic) inverter can be implemented with a 'digital transistor' as shown in Figure 1.37(a), using an SM resistor as collector load. These digital transistors, from Rohm, are available in the tiny 3 pin packages shown in Figure 1.29, with a variety of values for R_1 and R_2. For example, type *DTC144ExA* (where x is a code indicating which of the three packages of Figure 1.29

Advanced circuit techniques, components and concepts 53

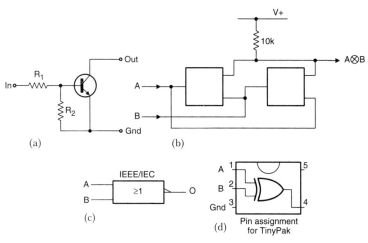

Figure 1.37 *Digital transistors, (a), from Rohm, are available in SOT23 packs (Figure 1.29), with a variety of values for R_1 and R_2. Two such transistors connected as in (b) give the inverse EXOR or exclusive NOR function. A single component solution is also possible, being readily available in CMOS as the NC7S86M5, (c) and (d), from National Semiconductor*

applies) is an NPN transistor where $R_1 = R_2 = 47K$. Adding another such transistor connected to the same collector load provides the NOR function, whilst connecting them as in Figure 1.37(b) gives the inverse EXOR or exclusive NOR function. With three separate components, this provides just about the most flexible layout possibilities that could be devised.

However, a single component solution is also possible, for nearly all the functions which are available in quad SSI packs are also available as singles in the SOT23-5 pack (one example has been illustrated already in Figure 1.33(a)). Suppose, for example, that an EXOR gate were required, this is readily available in CMOS as the *NC7S86M5*, see Figure 1.37(c) and (d), from National Semiconductor, along with AND, NAND, OR, NOR gates, etc. The device quoted operates from supplies of 2 to 6 V, sinks or sources 2 mA and has a propagation delay T_{pd} of 4.5 ns typical.

As well as the large packages of Figure 1.36, special-purpose digital ICs are available in the smaller packs discussed here. A good example is the *REG5608*, which is an 18 line SCSI (small computer systems interface) active terminator chip from Burr-Brown, Figure 1.38.

On-chip resistors and voltage regulator provide the prescribed SCSI bus termination, whilst adding only 2 pF per line, important for SCSI FAST-20 operation. All terminations can be disconnected from the bus with a single control line, the chip output lines then

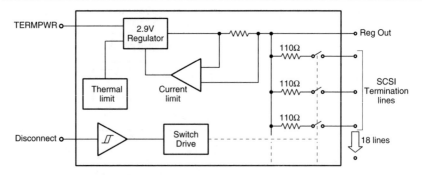

Figure 1.38 *The REG5608 is an 18 line SCSI (small computer systems interface) active terminator chip from Burr-Brown. On-chip resistors and voltage regulator provide the prescribed SCSI bus termination. A single control line open circuits all the terminations, important for 'hot socket' equipment plugging. The device is available in both 28 pin SOIC and fine pitch SSOP packages*

remaining high impedance with or without power applied, important for 'hot socket' equipment plugging. The device is available in both 28 pin SOIC and fine pitch SSOP packages.

Technical considerations

When using the very small types of components discussed above, a somewhat different approach is called for, compared with ICs in DIPs and other easily handled parts. The sheer practical difficulties of conventional breadboarding have already been mentioned. Consequently, with these very small parts, extensive circuit simulation to (hopefully) finalise the design, followed by going straight to PCB layout, is the usual order of the day. (In any case, if the circuit also involves one or more of the fine pin-pitch multi-pin devices, some of which are illustrated in Figure 1.36, then a PC layout will be required at the outset anyway.) Simulation is eased by the availability of Spice models for many of these devices; even if not, a simple model using just the input capacitance, first and second breakpoints and the output resistance may prove adequate. It is also useful to add a few strategically placed pads or TPHs – through plated holes – to provide testpoints for use in evaluation and debugging. This is safer than trying to probe pins which are spaced a millimetre or less apart.

Manufacturers face various problems producing very small parts. One concerns packaging, where the package dimensions may not be much larger than the basic silicon chip itself. For example, the *LT1078/9* and *LT1178/9* family of single-supply opamps in standard

DIP format from Linear Technology, with their low 55 µA, 21 µA supply current per opamp respectively, are justly popular. But the same devices in the surface-mount SO outline exhibit worse maximum input offset voltage V_{os}, and offset voltage drift. This is because the plastic surface-mount packages, in cooling, exert stress on the top and sides of the die, causing changes in the offset voltage. In response to this problem, Linear Technology has introduced the *LT2078/9, 2178/9* range. These new devices use a thin (approx. 50 micron) jelly-like coating, applied before encapsulation, to reduce stress on the top of the die, resulting in significantly better V_{os} and V_{os} drift.

Manufacturers also face problems with the marking of these very small parts. The capacitance value is marked on ATC ceramic chip capacitors, for example, in neat clear print, but which is so tiny it can only be read with the aid of a powerful eyeglass. IC designations tend to be quite long, so manufacturers are often obliged to use abbreviated codes to designate a part. For example, the SOT23-5 packaged *NC7S86M5* exclusive OR gate of Figure 1.37 is marked simply '7S86' on the top, whilst the similarly packaged *LMC7101BIM5X* opamp, also from National Semiconductor, is marked 'A00B'.

Figure 1.36 illustrates another point one should be aware of when using these devices – watch out for the mechanical dimensions. While the two 44 pin devices illustrated in Figure 1.36(c) and (d) look very similar, the pin pitch on the *ST44* in (c) is 0.8 mm, whilst that on the *ST44A* in (d) is 1.0 mm. Pin connections are another possible trap. The connections for a single opamp in the SOT23-5 package shown in Figure 1.30(a) are the commonest variety, used by a number of manufacturers. But some SOT23-5 opamps use pin 1 and 3 as inputs, with pin 2 ground, and the output on pin 4.

With today's densely packed boards, multilayer PC construction is the order of the day, usually with inner power planes and signal runs on the top and bottom planes. Interconnections between top and bottom planes, often used for mainly horizontal and vertical runs respectively, is by vias or TPHs, whilst 'blind' vias may be used for connections to or between inner layers. Unfortunately, the minimum pitch of conventional TPHs is greater than the pitch of the pins on many packages. So adjacent TPHs have to be staggered, taking up more board space, and negating some of the advantages of the very small packages. A more recent development, microvias, provides a solution, at a cost. These are so small that they can be located actually within the land area of each pin's pad, permitting much closer spacing of ICs.

Despite the extra considerations which applying these very small devices imposes, they can benefit the designer in many ways. For example, two single opamps in SOT23-5 packages occupy about half the board space of a dual opamp in an SO-8 pack. Additionally, even more space saving may accrue, due to the greater flexibility afforded by two separate packages. Each can be placed exactly where needed, minimising PC trace lengths. The problem of needing the odd gate, right across the other side of the board from a bespoke masked logic chip or ASIC containing all the other logic, has already been mentioned. Individual gates and buffers such as that in Figure 1.37 clearly supply the answer. But they have another use, no less important. They can be used to add a buffer to an output of the ASIC, found to be evidently overloaded at board evaluation stage, or even to implement a minor last minute logic change without the cost and delay penalty of having to redesign the ASIC – provided that at the layout stage, the designer took the precaution of leaving a spare scrap of board area here and there.

With all their advantages, tiny ICs, both analog and digital, are destined to play an increasingly important role in today's electronic world, where time to market is all important.

2 Audio

Low distortion audio frequency oscillators

Low distortion is a relative term. This article describes a simple oscillator design covering 20 Hz–20 kHz in three ranges, with distortion less than 0.05% and a tuning control with a linear scale.

Low distortion audio frequency oscillators

Readers of *EW+WW* have long shown a lively interest in high fidelity reproduction, dating from before the days of the Williamson amplifier. Since an early quasi-complementary design (Tobey and Dinsdale, 1961) appeared, many solid state high fidelity amplifier designs have featured in its pages, including those by Nelson-Jones and Linsley Hood. The evaluation of amplifier performance requires a low distortion AF oscillator, or rather two since intermodulation testing is essential nowadays. *EW+WW* has published many designs for these, including an early one by Rider. Particularly noteworthy is an APF (all-pass filter) based oscillator using a distortion cancelling technique (Rosens, 1982). This uses a Philips thermistor type 2322 634 32683, and the circuit achieves a very low THD (total harmonic distortion), namely <0.005% at 20 Hz and around 0.0002% at 1 kHz.

What sort of performance can be obtained without using a thermistor? The obvious and cheapest alternative amplitude stabilisation method is to use a diode limiter. This avoids using a specialised and expensive component; moreover it is aperiodic and so completely removes the annoying amplitude bounce often found in instruments using thermistor or AGC-loop stabilisation, when changing frequency. With diodes, a design based on the SVF (state variable filter) is preferable to the all-pass filter approach, since the former offers an inherent 12 dB octave roll-off at the lowpass output. This approach is a great help as the clipping will produce all the odd harmonics one expects to find in a squarewave: in contrast, the small amount of distortion produced by a thermistor is almost pure third harmonic. This makes distortion cancellation by outphasing in the design very effective even in an APF-based design. Incidentally, in an SVF-based oscillator the desirable feature of a linear scale is easily engineered.

Figure 2.1 shows an SVF-based oscillator which, with the component values shown, operates at 1.59 kHz. V_1 (the voltage at the lowpass output, OUTPUT 1) was 5.3 V pp (volts peak-to-peak). The Q of this two-pole filter is $R_5/3R_4 = 11$ with the values shown. This modest value of Q was chosen deliberately, to enable the effect of circuit changes on the output distortion to be easily seen. A Q of 30 would be quite usable and is indeed used in a 20 Hz–20 kHz 0.04% distortion SVF-based sinewave generator currently in production (the Linstead G3 Sine, Triangle, Square Audio Signal Generator, manufactured by Masterswitch Ltd).

The circuit operates as follows. If an input were applied via a resistor to the inverting input of IC_1A, the bandpass output BP would be in phase with it at the frequency where the gain of each of the integrators is unity. So if the bandpass output is taken and clipped (an aperiodic operation introducing no phase shift) to a squarewave, the

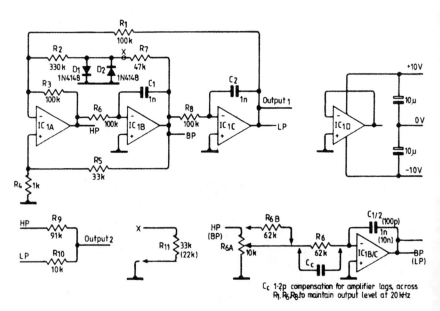

Figure 2.1 (a) A 1.59 kHz oscillator circuit based on a state variable filter. (A detailed explanation of the SV filter's operation is given in Hickman (1993).) (b) Modifications give a nil net third harmonic at OUTPUT 2

Audio 59

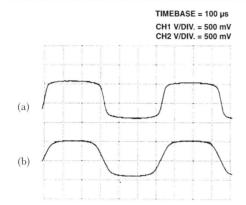

Figure 2.2 (a) Smoothness of square waveform across V_d. (b) Connecting R_{11}, 33 kΩ in parallel with the diodes results in more gently sloping sides

latter can be used as a fixed level excitation input to the filter. At the filter *LP* (lowpass) output, the fundamental appears amplified by the Q factor while the harmonics are reduced due to the filter's 12 dB/octave roll-off. Figure 2.2(a) shows the waveform across the diodes V_d, which is 0.96 V pp and a rough approximation to a squarewave. Assuming that the fundamental component is about 1.4 V pp, the output should be $1.4Q(R_3/R_2) = 4.7$ V pp, not so very different from that observed. If the squarewave input to the filter were ideal, the amplitude of the third harmonic component would be one-third that of the fundamental. At the filter's output the third harmonic will have been attenuated by a factor of 3 in each of the integrators, whilst the fundamental will be amplified by the Q factor of 11. The third harmonic should therefore be one-third of one-ninth of one-eleventh of the fundamental, or 0.34%. The fifth and higher harmonics will be substantially lower, due to the filter's 12 dB/octave roll-off, so the expected distortion is approximately 0.34% and all third harmonic. In fact the measured distortion is 0.2%, due to the smoothish nature of the squarewave of Figure 2.2(a).

With the SVF, the signal at the *LP* output is always in antiphase to the signal at the *HP* (highpass) output – another advantage over the all-pass design in this application, as it simplifies the outphasing of the distortion. This is achieved by making the circuit operate as a second-order elliptic filter at the same time as an oscillator, as follows. The circuit oscillates at the frequency of the peak of the *BP* (bandpass) response and at this frequency the gain of the two integrators is unity, so the fundamental component of the output has the same amplitude at the *HP*, *BP* and *LP* outputs whereas the third harmonic is attenuated by a factor of 3 in each of the integrators. So by combining one-ninth of the HP signal with the LP signal to give V_2 as OUTPUT 2 as in Figure 2.1(b), the net third harmonic at OUTPUT 2 is nil: we have placed a zero in the filter's response at three times the frequency of the BP peak response. Meanwhile, V_2 has been reduced by about 1 dB by the partial outphasing of the

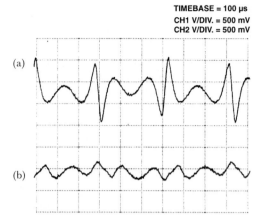

Figure 2.3 (a) Residual distortion component in OUTPUT 2. Fifth harmonic predominates. (b) Residual after connecting R_{11}

wanted fundamental output. The absence of third harmonic is clear in Figure 2.3(a), showing the residual distortion component in OUTPUT 2. A count of the peaks of the waveform shows that fifth harmonic predominates, but there is a rapid spiky reversal, corresponding to the steep sides of the waveform shown in Figure 2.2(a). This shows that the higher harmonic components of Figure 2.2(a) are far from negligible. Nevertheless, the distortion is reduced from 0.2% to 0.095%, a useful if not spectacular improvement. Clearly matters would be improved if the clipping were gentler, the problem being that the BP output amplitude into the limiter is ±2.6 V peak, whereas the diodes clip at only 0.5 V.

Connecting R_{11}, 33 kΩ, in parallel with the diodes moves the point at which clipping occurs up nearer the peak of the waveform, resulting in more gently sloping sides, Figure 2.2(b). The level of the fundamental component is little affected, the output falling only by 0.5 dB. The residual is then as in Figure 2.3(b), the distortion is 0.034% and is visibly almost pure fifth harmonic. If R_{11} is further reduced to 22 kΩ, with a further 0.5 dB drop in output, the distortion falls to 0.018%. Thus with R_5 raised to provide an operating Q of 30, a distortion level of 0.006% could be expected, a very creditable performance to such a simple circuit. Indeed, one would need to consider using a lower distortion opamp such as the *NE5532* used in Rosens (1982), the *TL084* used having a typical total harmonic distortion of 0.003% up to 10 kHz.

The performance is still substantially inferior to that in Rosens (1982) and the reason is not far to find. With a second-order filter, we can only engineer a zero in the stopband response at one frequency. So although the third harmonic can be outphased, the filter's response rises again beyond that frequency. Consequently, the arrangement actually makes the fifth harmonic level in the output worse. This is where the thermistor scores, any small variation in its resistance over a cycle at the operating frequency resulting in almost no harmonic distortion other than third. It is tempting to speculate that by further waveform shaping in the non-linear network, one could restrict the harmonic distortion in the drive signal applied via R_2 to the filter's input to fifth and higher harmonics. The outphasing could then be modified (R_9 becoming 250 kΩ) to suppress the fifth harmonic. It is true that this would worsen the seventh harmonic in the output, but not by nearly as much since the ratio of 5^2 to 7^2 is much less than the ratio of 3^2 to 5^2.

However, although this is doubtless possible in a fixed frequency oscillator, the necessary settings would almost certainly be too critical to hold in a tunable oscillator covering 20 Hz to 20 kHz. Incidentally, the linear scale is organised as shown in Figure 2.1(b). With $R_6 = 62$ kΩ, the frequency range is 2 kHz down to zero. At mid-track (1 kHz) the loading of R_6 on R_{6A} causes the output frequency to be a little too low, since R_6 now sees the pot as a 2K5 source impedance instead of zero at max. and min. This parabolic error can be changed to a much smaller cubic one by connecting R_{6B} (=R_6) from the wiper to the top of track, giving zero error at 1 kHz.

References

Hickman, I. (1993) *Analog Electronics*, Heinemann Newnes, Oxford.
Rosens, R. (1982) Phase-shifting oscillator. *Wireless World*, February, 38–41.
Tobey, R. and Dinsdale, J. (1961) Transistor audio power amplifier. *Wireless World*, November, 565–570.

Free phase oscillators

To provide the richness of sound and convincing build-up of the chorus of a real pipe organ, many electronic organ constructors believe that there is no substitute for an independent oscillator tone-generator per note. For such a design to be realisable, an oscillator design combining simplicity, cheapness and very high stability is needed. This article looks at one such design.

Notes on free phasing

Introduction

Practical analog circuit design is fraught with snags, compromises and difficulties at every turn. These are well illustrated by the subject of this article – keyed tone generators, such as might be used in the two-tone alarm generator of an HF radio telephone or a hundred other applications. One of these applications is as tone sources in an electronic organ, or rather in one class of electronic organs, for there are a number of distinct approaches to design of these, each with its own advantages and disadvantages. The main varieties are divider organs and free phase organs. The former use a digital 'top octave generator' to produce the 12 semitones of the equal tempered scale, all the intervals being, if not exact, at least very close, and of course 'set in concrete'. Each semitone output is applied to a binary divider such as the seven stage *CD4024* to provide the lower octaves. Advantages of this approach include cheapness and simplicity (though top octave generators are not as easy to obtain as they once were) and an organ which is always in tune, but there are a number of snags as well. With all 12 semitones of seven or more octaves available all the time, each individual note has to be passed when the corresponding key is pressed, or else blocked, by its own keying circuit. It is difficult to obtain sufficient attenuation when notes are not supposed to be sounding, leading to a residual background noise aptly described by the term 'beehive'. Also, squarewaves contain no even harmonics, so some combining of different octave outputs for each note is necessary if a convincing variety of pipe-like sounds (especially open diapasons) is to be achieved, adding to the complexity. However, for anyone wanting at least an approach to the richness of sound provided by a real pipe organ, a major snag is the use of dividers to provide the various octave pitches. For example, if whilst sounding middle C an octave coupler is activated, then C′ (the C one octave above) will also start to sound. But since C was obtained by dividing C′ by two in the first place, the two notes are locked together, the octave is too perfect. In fact, all you have done is to change the harmonic content of C: if you didn't hear the two notes starting to sound at different times, you would never know that there were supposed to be two separate notes sounding. For this reason more than any other there is still a lively interest in 'free phase' designs, despite the availability of palliatives such as phase modulated delay lines which try to 'unlock' the various octaves.

An oscillator for free phase designs

A true free phase organ needs a separate oscillator for each note of the rank (or for half that number using an ingenious scheme for sharing one oscillator for each adjacent pair of semitones, on the premise that normal music does not require both to sound at once (Ref. 1). For example, on the usual five octave keyboard a flute stop would have 61 generators. The usual arrangement is $C_{,,}$ – two octaves below middle C – to C‴, three octaves above (whereas unfortunately the keyboard I have in stock against the day I actually get around to building an organ covers five octaves F to F). On an '8 foot rank' (so called because that is the length of the lowest pitch open flue pipe of the five octaves), middle C sounds at that pitch, whereas on a 4 foot rank, middle C would sound the note C′, and on a 16 foot rank, the note C,. To simulate the richness of a pipe organ, several ranks of generators are needed, corresponding to the different stops on a real organ. So clearly economy is a prime consideration in choosing an oscillator design, but equally important is stability. With 61 individual independent generators per rank, retuning would otherwise be an endless chore, unlike the case with a top-octave/divider approach.

In the past, many electronic organ builders have used *LC* oscillators, the inductor using a gapped laminated core. This type of oscillator has the advantage of not needing a separate keying circuit; it performs its own keying function by switching the supply to the maintaining transistor. The output is taken from a point in the circuit where there is no change in dc level between the on and off states, avoiding keying thumps, whilst the smooth build-up and decay of the amplitude avoids the slightest suggestion of 'keyclicks', which plague many other designs of keyed oscillators and keying circuits. Many such ranks are still in use, but the size and cost of using *LC* oscillators provides a strong incentive to seek alternative designs.

I therefore set myself the task of designing a cheap, simple, keyed oscillator (requiring no separate keying circuit) and requiring only an SPNO (single pole normally open) switch for each key contact – some published designs require, at each key, one changeover contact plus two normally open contacts. An SPNO contact is preferred to an SPNC contact, since the worst that dust can then do is to prevent a note from sounding when played, whereas with a normally closed contact, it can cause a note to become 'stuck on', known in organists' parlance as a cypher.

One of the simplest possible oscillators consists of a Wien bridge and an opamp, see Figure 2.4.

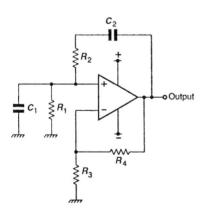

Figure 2.4 *Simple audio oscillator or tone generator, based upon the Wien bridge*

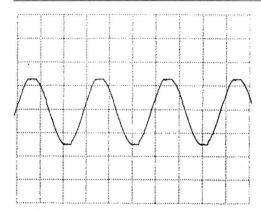

Figure 2.5 *Showing the output of an oscillator to the design of Figure 2.4*

The attenuation from the opamp output to its NI (non-inverting) input via $R_1 R_2 C_1 C_2$ is infinite at 0 Hz and infinite frequency, and a minimum of a factor of three at the frequency given by $f = 1/(2\pi RC)$, if $R_1 = R_2 = R$ and $C_1 = C_2 = C$. This forms the narrowband PFB (positive feedback) path. If the attenuation in the broadband NFB (negative feedback) branch via R_3 and R_4 is less than 3:1 the circuit will not oscillate, but if it is equal to (or in practice, because the gain of an opamp is finite, slightly greater than) 3:1, then the circuit will oscillate. With no special amplitude stabilising measures, the amplitude of the oscillation will build up until limited by the output hitting the supply rails, causing little distortion if the PFB signal at the NI input barely exceeds the NFB at the inverting input, see Figure 2.5. Surprisingly, using the circuit shown, with an LM324 opamp (the cheapest quad opamp you are likely to find), there is no audible change in pitch as the supply rails are varied from ±3 to ±15 V.

To make a practical organ tone generator, some means of tuning is required, and this is by no means straightforward. Varying any one of R_1, R_2, C_1 or C_2 will change the frequency, but will also change the attenuation in the PFB path, causing the oscillation to stop, or alternatively to limit so hard as to verge on a squarewave, depending on which way the attenuation changes. A two-gang resistor will do the job, but is hardly practicable on a one per note basis. But fortunately, as is so often the case in analog circuit design, where only a small parameter change is required a little ingenuity can provide the solution, Figure 2.6. If the reactance of the capacitor C_1 at the operating frequency is ten times the track resistance of the potentiometer, the voltage at B will be only 0.5% smaller than at A even though the voltage across the resistor will be one-tenth of that across the capacitor, since these voltages are in quadrature. However, as the wiper of the pot is moved from A towards B, additional phase lag is introduced onto the signal fed to the opamp's NI terminal. To compensate for this, maintaining zero phase shift from the opamp's output to its NI input, the frequency must fall. Due to the low Q of the RC network (its $Q = 1/3$), a small change in phase shift causes a

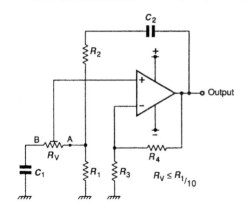

Figure 2.6 The addition of trimmer potentiometer RV permits tuning of the oscillator without changing the attenuation via the Wien network, provided the resistance of the pot's track does not exceed one-tenth of the reactance of C_1.

much larger compensating change in frequency than would be the case with an LC circuit. At the operating frequency, the reactance of C_1 equals R_1, so in Figure 2.6, the track resistance of the pot should not exceed 10K – this provides almost three semitones tuning range, while a 4K7 pot provides over one semitone.

My lab notebook, Volume 4, records that I developed this circuit in 1982, but I know that it has been independently derived by others (Ref. 2). It has a further advantage in that the wiper of the pot feeds an opamp input, i.e. a high impedance. Except in the case of wirewound types, the resistance from one end of a pot to the wiper plus that from the wiper to the other end, exceeds the end-to-end track resistance, due to wiper contact resistance. The contact resistance is relatively less stable than the track resistance, so tuning by making part of R_1 or R_2 a pot would be impracticable on stability grounds, quite apart from the incidental change in loop gain. As it is, C_1, C_2 can be polystyrene types, available in E12 values at 1% or more cheaply 2.5% selection tolerance. The resistors should all be metal film types: nowadays these are little more expensive than carbon film, and like many of his colleagues, the author has changed over to these as stock items. Using polystyrene capacitors and metal film resistors, the long-term stability of the oscillators should be adequate to ensure that only occasional retuning is necessary. Over the temperature range 20°C to 60°C, the breadboard circuit exhibited a tempco of –0.02%/°C, using polycarbonate capacitors. The frequency shift with change of ambient temperature can be expected to be (for all practical purposes) the same for all notes, provided of course that the capacitors used all have the same type of dielectric.

Having arrived at a stable, tuneable oscillator, it remained to add a keying facility, which can be achieved by altering the ratio of R_3 and R_4. This has to be effected by the key contact, but the latter cannot be used to modify the component values directly, if – as is likely – it is required

66 Analog circuits cookbook

to add octave and suboctave couplers. These, when activated, sound the note an octave above, and/or an octave below each note played. This increases the richness of sound and, because of the inevitable slight departure from exact octaves when using individual generators, creates a desirable chorus effect just as in a real pipe organ. Thus the key switches should simply key a dc control signal, instructing the generator to sound when the corresponding key is depressed. The circuit itself will be controlled by an electronic switch. CMOS switches are cheap and readily available and, like the *LM324* opamp, come four to a pack, e.g. the *CD4016*, so this device was selected.

Figure 2.7(a) shows such a keyed oscillator whilst Figure 2.7(b), upper trace, shows the output waveform, which is basically sinusoidal

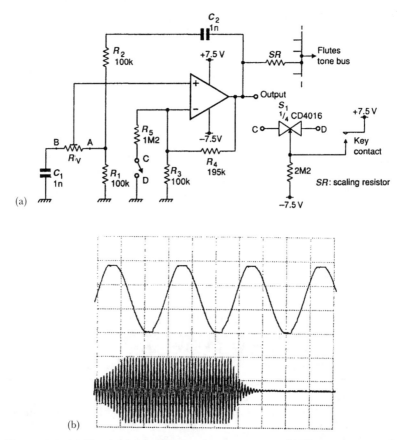

Figure 2.7 (a) Circuit of a keyed sinewave generator. (b) The output waveform is basically sinusoidal, suitable for use directly for stops of the flute family, upper trace. The starting and ending transients are smooth and free from any incidental dc shift, lower trace

and hence suitable for use directly as the basis of stops of the flute family. Figure 2.7(b), lower trace, shows the starting and ending transients, which are clean and smooth, and with no associated dc level shifts, giving complete freedom from key clicks and thumps respectively. The note sounds when R_5 is grounded via S_1, one section of a *CD4016*. In view of the supply voltage rating of this device, the circuit is run on ±7 V rails instead of the more usual ±12 or ±15 V. R_6 normally holds the control pin of S_1 at −7 V, the key contact raising this to +7 V to sound the note. The rate of build-up of the tone depends on how much greater than 3:1 is the attenuation from the opamp's output back to its inverting input when the key is depressed, whilst the rate of decay is set by how much less than 3:1 when the key is released. Thus by suitable selection of R_3, R_4 and R_5, the attack and decay times can be separately adjusted. For although Figure 2.7(a) behaves like a high Q tuned circuit, this is only because the feedback is just too much or too little to allow it to oscillate. Where the frequency determining network has a high Q in its own right, e.g. an *LC* oscillator, the build-up transient will generally be as fast as the decay – or faster if the maintaining circuit is heavily overcoupled.

Creating other tone colours

While a near sinewave is fine for flute-type stops, waveforms with higher harmonic content are needed to simulate many other pipe sounds. A near squarewave, with its absence of even order harmonics, is ideal for stops of the clarinet family, and Figure 2.8(a) shows a simple add-on circuit to provide it; of course one per note is required. Figure 2.8(b), lower trace, shows the 'squarewave', compared with the input sinewave driving it, upper trace. Due to its rather smooth shape, the harmonics, especially the very high ones, roll off rather faster than a true squarewave, but it sounds very acceptable. Figure 2.8(c) shows the ending transient, which – due to the limiting action of the diodes – is extended compared with the sinewave. In practice, this is of no consequence, provided it is smooth, well controlled and free from clicks or thumps, as the ear is much less sensitive to the end of a note than it is to its beginning.

For other types of sound, for example open diapasons, some second harmonic is essential. Stopped diapason pipes, being a quarter of a wavelength long, are an exception, but even these, if of large square cross-section tend to show some second harmonic. Figure 2.9(a) shows an interesting shaper circuit, originally published in an American magazine, and modified here with suitable component values for the available drive voltage. Figure 2.9(b) shows the output

68 Analog circuits cookbook

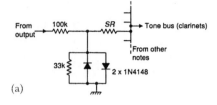

(a)

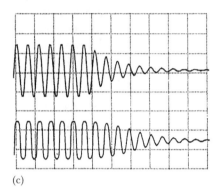

(c)

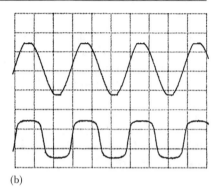

(b)

Figure 2.8 *(a) A simple clipper circuit provides an approximation to a squarewave. (b) Comparing the 'squarewave', lower trace, with the input sinewave. (c) Due to the limiting action of the diodes, the ending transient of the squarewave output is extended compared to that of the sinewave*

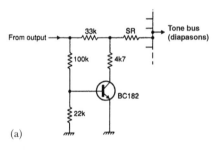

(a)

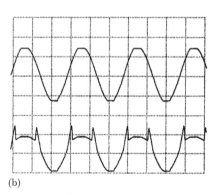

(b)

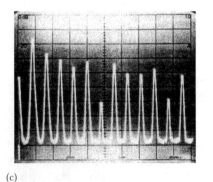

(c)

Figure 2.9 *(a) A circuit for adding second (and other) harmonics to the sinewave. (b) The output of the above circuit, lower trace, compared with the sinewave input, upper trace. (c) Showing the fundamental at about 1.7 kHz, the second harmonic about 10 dB down (about right for an open diapason), and many other harmonics. (10 dB/div. vertical, 2 kHz/div. horizontal, span 0–20 kHz)*

voltage, lower trace, compared with the input sinewave, upper trace. The harmonic content of the waveform of Figure 2.9(b) is shown in Figure 2.9(c). Experimentation with the relative values of the four resistors enables a wide variety of waveshapes, and hence of harmonic content, to be achieved. However, in the process of introducing even harmonics, the circuit reduces the area under positive-going half cycles more than under the negative-going ones. This means that it introduces a small dc component, which results in an offset at the keyed output relative to ground when sounding. The result is a slight tendency to produce keying thump, mitigated somewhat by the fact that the driving sinewave builds up and dies away gradually. This effect is found in nearly all schemes for introducing second harmonic, and the thump can be largely suppressed by passing the output through a highpass filter. The filter need not be provided on a one-per-note basis, but on the other hand one per rank cannot be effective over the whole keyboard. The Figure 2.9(b)-type tone generator outputs can therefore be combined on a per octave basis, passed through an appropriate highpass filter and the five filter outputs combined for feeding to further voicing and tone forming filters. If passed through a highpass circuit providing attenuation of the fundamental relative to the harmonics, a sound like a really fiery reed stop results.

By these means, three different stop types can be derived from a single rank of generators, but of course in no way does this make it equivalent to three independent ranks. Drawing two of the three stops together simply changes the harmonic content of a note. It therefore contributes nothing to the chorus effect, whereas with two different speaking stops drawn on a pipe organ, two different pipes sound for each note. Nevertheless, it is convenient to have three different tone colours available, even if drawing them in different combinations merely provides further different tone colours. In particular, one output can be voiced as a very loud stop and another as a quiet one: if the loud one were drawn the quiet one would not be heard anyway, even on a real pipe organ.

Cutting the cost and complexity

However simple the tone generator, the requirement for one per note per rank means a lot of kit is needed. The Ref. 1 scheme of sharing a generator between two adjacent semitones is therefore very attractive, but that circuit used a relaxation oscillator. But changing the pitch of a Wien bridge oscillator is not so simple as pulling the frequency of a relaxation oscillator. This is because, as noted earlier, whilst changing either R_1 or R_2 alone will change the frequency, it will

70 Analog circuits cookbook

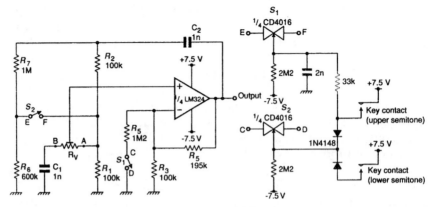

Figure 2.10 *The circuit modified to sound either of two adjacent semitones, according to which key is pressed. The addition of both R_6 and R_7 keeps the loop gain the same when S_2 is closed, leaving the amount of clipping at the rails the same for either semitone (see Figure 2.11(a))*

also change the required ratio of R_3 and R_4. What is needed is a way of simultaneously changing both R_1 and R_2, using – for economy – just a single pole switch, such as a single section of a *CD4016*. Here again, as the parameter change required is a small percentage – one equal tempered semitone represents a 5.9% change in frequency – a little ingenuity can supply the answer, Figure 2.10.

Whilst the two additional resistors connected to switch S_2 will marginally increase the frequency of oscillation when S_2 is open, values can be found which will cause a further increase of exactly a semitone in pitch when it is closed, without changing the PFB level. Thus the degree of clipping is unchanged (compare the two semitone outputs in Figure 2.11(a)), leaving the harmonic content virtually unchanged, Figure 2.11(b). Here, the semitone frequency separation of the two fundamentals is only just visible, but the separation becomes two semitones or about 12% at the second harmonic, and so on in proportion to the order of the harmonic. The starting and ending transients of the upper semitone are also unchanged, due to circuit arrangement maintaining the same degree of clipping for both semitones, Figure 2.11(c).

For the purposes of experimentation the actual frequencies were regarded as unimportant, the semitone shift being the essence of the exercise, but the two notes – in the region of 1700 Hz – correspond roughly to A″ and A″ flat. There is a small effect on the accuracy of the semitone change, depending on the setting of the tuning potentiometer. This amounts to a few cents more or less than a semitone with the tuning potentiometer at one extreme end of its

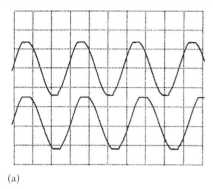 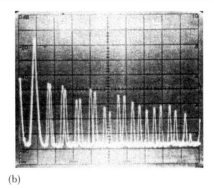

(a) (b)

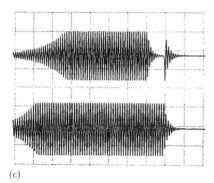

(c)

Figure 2.11 (a) The two sinewave outputs, a semitone apart. (b) As a consequence, the amplitude and harmonic content of the circuit's sinewave output is virtually the same for both semitones. (10 dB/div. vertical, 2 kHz/div. horizontal, span 0–20 kHz.) (c) Delaying the removal of the semitone pitch change control signal to avoid chirp on end transient of the squarewave output when sounding the upper tone causes a hiccup in the ending transient of the upper tone sinewave output, audible as a slight key click

range or the other, where one cent represents one-hundredth of a semitone.

The two diodes in Figure 2.10 are arranged such that either of the two adjacent semitone keys will close S_1, causing the note to sound, but only when the key for the upper note is pressed will S_2 be closed, giving the higher of the two pitches. If both keys are pressed at once, the upper semitone sounds, unlike some shared note schemes, where accidentally pressing both keys together causes a totally different, unrelated note to sound. With the 2n capacitor (shown feint) absent, the pitch will revert to the lower semitone immediately the upper semitone key is released, causing the tail of the note to be at the lower semitone frequency. Strangely, this results in but the barest trace of keyclick on the sinewave output, presumably because of the rapid decay of the tone, Figure 2.7(b). However, the decay of the squarewave output is much slower, due to the limiting action of the diodes, and this is clearly visible in Figure 2.8(c). Hence on the squarewave output, the pitch change during the ending transient of the upper semitone gives a much more obtrusive keyclick. The 2n capacitor suppresses this by delaying the return to the lower pitch when the key is released. The resistor (shown feint) is necessary to control the capacitor charging current, otherwise a keyclick appears at the beginning of the upper semitone squarewave output.

Unfortunately, whilst the bracketed components suppress any keyclick on either semitone on the squarewave output, they create a very audible keyclick on release of the upper semitone sinewave output. This is caused by charge injection in the switch circuit S_2, from the control input to that section of the *CD4016*. With the capacitor delaying the opening of the switch, it now occurs when the sinewave has all but died away, and as the switch is connected directly to the opamp's NI input, it shock excites the oscillator into ringing – visible on the upper trace (upper semitone) in Figure 2.11(c). By comparison, the lower semitone sinewave output is of course unaffected, lower trace.

Charge injection in electronic switches is a well-known phenomenon, and in later designs of switch ICs it has been much reduced, but these would be too expensive in the numbers required for this application. Clearly there is scope for further development here, for example the capacitor at the control input of S_2 could be grounded not directly, but via another section of the *CD4016*. This additional section would be switched on when squarewave was selected, but not for sinewave. All the additional switch sections would have their control inputs connected together and controlled by the stop switches, being on for clarinet (squarewave) type stops but off for flutes (sinewaves).

This article has concentrated on the basic per-note (or per pair of notes) tone generator, but a word on controlling the generators from the keyboard will not be amiss. For a very simple organ of just one rank, the key switches can control the S_1 for each note directly, and the S_2 – if using the shared generator scheme – via diodes as in Figure 2.10(a). If it is desired to incorporate octave and suboctave couplers, this can be achieved with the addition of extra diodes and resistors, but the complexity increases alarmingly, especially with the shared generator scheme. It increases further if it is desired to have two or more ranks of generators with the option of sounding these at different pitches, so for all but the least ambitious designs, some other scheme is called for. A microcontroller can be employed to scan the keyboard and set or clear latches controlling S_1 (and S_2 if used), in accordance with the stops drawn. But a simpler approach is to employ one of the variations on the multiplex scheme, which has been described many times in the literature, e.g. Refs 3 and 4. A version of the scheme has also been described in these pages.

References

1. Asbery, Dr J.H. (1994) Shared note F-P oscillator. *Electronic Organ Magazine* (Journal of the Electronic Organ Constructors' Society), No. 154, December, pp. 14, 15.
2. Asbery, Dr J.H. (1992) A free phase organ. *E.O.M.*, No. 145, March, pp. 8–13.
3. *E.O.M.*, Organ Notes (No. 10), Dr David Ryder, November, 1981, No. 98, p. 12.
4. Hawkins, T. (1992) Experimenting with multiplex. *E.O.M.*, No. 146, July, pp. 12–15.

Externalising the sound

Listening to music through headphones has several advantages, perhaps the main being that you can have it as loud as you like without disturbing the neighbours, or even the rest of the family. But the main disadvantage is perhaps that the music sounds as though it is inside your head. Many years ago I was told by a colleague that this is because there is no differential change in the phase of the signals reaching the ears when the head is turned. Normally, there would be, this being the mechanism that enables one to tell the direction that a sound is coming from. I had long wanted to check out whether adding delays to the signals to the left and right earpieces, delays which varied whenever the head was turned, could 'externalise' the sound. But the opportunity to do so had not arisen. Doubtless the experiment has been performed before, but that is no reason for not trying it for oneself, and in any case it would surely provide some interesting design problems along the way.

Music in mind

Solution looking for problem

Recently I saw an advertisement for a miniature all-solid state gyro; here surely was a solution in search of a problem. One of the uses envisaged by the manufacturer is in automobile navigation systems, but clearly there are many others – the device would be a fascinating component to play with. Being fortunate enough to acquire a sample, here was an opportunity to try out the aforementioned psychoacoustic experiment. The gyro could be used to sense rotation of the head, and this signal used to adjust the delays in the left and right channels.

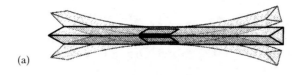

(a)

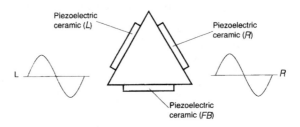

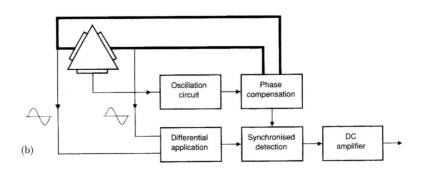

(b)

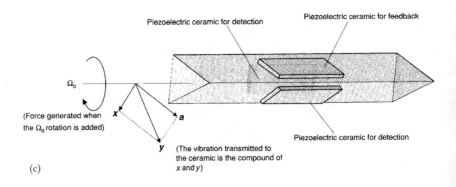

(c)

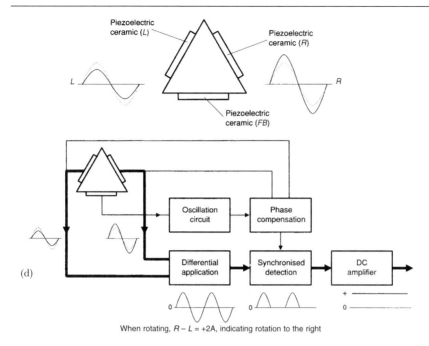

Figure 2.12 (a) The Murata piezoelectric vibrating gyroscope uses a triangular prism, maintained in a flexure-mode vibration. (b) All three electrodes, one on each face, are used to maintain oscillation, whilst two are also used to pick off any differential voltage due to rotation. (c) When rotation about the longitudinal axis occurs, the force transmitted to the prism contains an extra component 'a'. (d) This results in a corresponding differential voltage between the detection electrodes, proportional to the rate of rotation

The first step was to learn a little about the piezo-vibrating gyroscope. This uses a triangular prism of elinvar metal (to which are attached piezoelectric transducers), maintained in a flexural mode oscillation, by an oscillator operating at its resonant frequency, Figure 2.12(a). The vibration is maintained by a set of three electrodes, Figure 2.12(b), two of which are also used as sensors. When the unit is rotated about the longitudinal axis of the prism, an additional component of force is applied to the piezoelectric material, Figure 2.12(c). This results in a differential component in the voltage at the two detection electrodes as in Figure 2.12(d): this is picked off and synchronously detected, filtered and smoothed, providing a voltage proportional to the rate of change of direction.

Figure 2.13 shows an application circuit which appears on the manufacturer's data sheet for the device. Note that the signal output is ac coupled: this is to allow for a possible standing offset between

76 Analog circuits cookbook

the signal output and the reference voltage to which it relates, and in particular for temperature variation of this offset (there is also a temperature coefficient of the nominal 1.11 mV/°/sec scale factor). In an automotive navigation system, it is assumed that the vehicle will return to a straight-line course after each turn before the highpass filter introduces too much signal loss, but if you were to drive round and round a roundabout the system might presumably lose track of the vehicle's direction. For since the device produces a signal output (relative to the reference) which indicates the rate of turn, this signal must be integrated to obtain an output giving the actual direction of travel. However, it is possible to engineer a 3 dB corner frequency much lower than the 0.3 Hz shown in Figure 2.13, avoiding this problem whilst still blocking the much slower variations in output offset due to temperature variations.

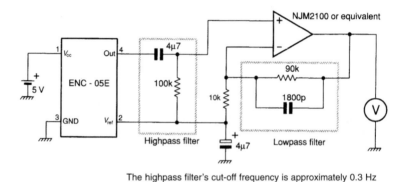

The highpass filter's cut-off frequency is approximately 0.3 Hz
The lowpass filter's cut-off frequency is approximately 1 kHz

Figure 2.13 *Sample amplifier circuit from the ENC-05E A1 solid state gyro data sheet. (Note that the base diagram shown is confusing; V_{ref} is actually on the same side of the device as V_{cc})*

For the purposes of the psychoacoustic experiment, the gyro would be fixed to the headband of a pair of earphones, to detect head movements. So the gyro was mounted on a small piece of 'VERO' 0.1 inch matrix copper strip board, with a couple of metres of screened lead for the signal and earth connections, and two other wires, for the +5 V supply to the unit and its reference output V_{ref}. The signal output was passed through an ac coupling with a time constant of 300 seconds, giving an LF cut-off frequency of about 0.0005 Hz. This is arranged as shown in Figure 2.14, which shows the gyro output applied through a 100K plus 10n lowpass filter to further suppress switching ripple in the signal output, to the input of a unity gain buffer stage A_1. The 10 MΩ resistor at the NI (non-inverting) input

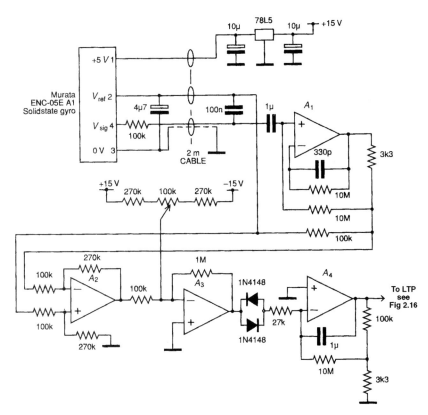

Figure 2.14 *Circuit diagram of the gyro output signal-conditioning stages, plus the integrator which turns the rate-of-rotation signal into an azimuth position signal*

of A_1 is returned not to V_{ref}, but to a point at 97% of A_1's output. This effectively multiplies its value by a factor of 30 giving, in conjunction with the 1 μF capacitor, a time constant of 300 seconds.

A *TLE2064* quad opamp was chosen for A_1 (also used for A_2–A_4), on account of its low bias current I_b of 3 pA and offset current I_o of 1 pA – both typical values, at 25°C. The buffered high- and lowpassed signal output and the reference output were applied to A_2, connected as a bridge amplifier providing rejection of the common-mode reference voltage. Its output is thus ground referenced, adequate common mode rejection being obtained due to the use of 1% metal film 100K and 270K resistors.

A_2 provides a gain of ×2.7 and a further gain of ×10 is raised in A_3, at which stage an offset adjustment is introduced, to allow for offsets in A_1 and A_2. In practice, on switch-on it was necessary to temporarily short the 10 MΩ resistor at the NI input of A_1, to avoid a very long wait for the dc conditions to settle. On removing the short, there was still an offset due to I_b flowing in 10 MΩ rather than a short circuit. So a 10 MΩ resistor was included in the inverting input also, bypassed by a 330 pF capacitor, to maintain stability. A normally open two pole switch was used to short both 10 MΩ resistors at switch-on, to allow for settling. Even so, drift of the output of A_1 was still experienced, so finally the 1 μF capacitor and the resistors were removed, and A_1 reconnected as a simple dc coupled unity gain buffer. The offset between the signal and reference outputs of the gyro turned out to be only a few millivolts, and could thus be nulled with the offset adjustment at A_3's input. As ambient temperature changes in a domestic environment are small and slow-acting, this proved acceptable for the purposes of this experiment.

The output of A_3 was integrated, to obtain the absolute rotary position of the headphones. But here there is a problem; integrators have an annoying but unavoidable habit of heading off, over the long term, to one or other of the supply rails, since in practice, the input voltage will never remain exactly at zero. The solution used was twofold. First, when the listener's head is stationary, giving no output from the gyro and hence none from A_3, the 27 kΩ resistor at the integrator's input is effectively disconnected by the two diodes. Furthermore, to prevent the integrator from integrating its own input bias current, a 3000 MΩ resistor was connected across the 1 μF integrating capacitor. Actually, a 10 MΩ resistor was used, but since only one-thirtieth of the integrator's output is applied to it, its effect is that of a 3000 MΩ resistor. This means that, in the absence of head movements, the 'sound stage' will over a period of many minutes revert to straight ahead. This is where it should be of course, if it be assumed that the listener will not want to spend long periods with his head cocked uncomfortably to one side or the other.

Note that considerable gain has been used ahead of the integrator, so that even comparatively small, slow movements of the head will produce a large enough output from A_3 to turn on one or other diode, effectively reconnecting the 27K resistor at the integrator's input.

Checking the delays

The output of the integrator, indicating the rotational position of a listener's head, was used to control the relative time delay of the sounds reaching the ears. To find what this should be, some simple measurements and calculations were needed. With the aid of a ruler and a mirror, I determined that my ears were about 14 cm apart. Thus, when the head is turned through an angle of 45° to left or right,

Audio 79

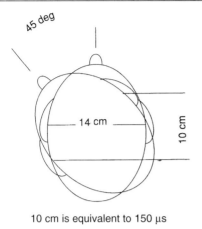

10 cm is equivalent to 150 µs

Figure 2.15 *Showing the differential delay to binaural sounds as a function of head rotation*

one ear moves to a position, in the fore–aft direction, 10 cm ahead of the other. So each channel needs to be able to produce a delay equivalent to ±5 cm, or (given the speed of sound is about 1100 feet per second) ±150 µs, Figure 2.15.

BBDs (bucket brigade devices) were used to produce a delay in the signal to each earphone, with the delay being varied by variation of the clock frequency used to drive the BBDs. The 1024 stage Panasonic BBDs type *MN3207* were each driven by a matching *MN3102* CMOS clock generator/ driver. This contains a string of inverters which are usually used in conjunction with an external R and C, setting the clock frequency. For this application, the R and C were omitted, and the first inverter driven by an externally generated clock. The two clock generator/ drivers were driven by two VCOs (voltage controlled oscillators). These in turn were controlled by an LTP (long tailed pair) driven from the output of the integrator in Figure 2.14.

Initially, an elegant VCO using an OTA (operational transconductance amplifier) and a *TL08x* opamp was designed and tested. This had the advantage of providing a unity mark/space ratio independent of output frequency, but was abandoned as it would not run fast enough – the drive to the clock generator/driver chips has to be at twice their clock output frequency. So a pair of simple VCO circuits, using two sections of a *CD4093* quad two-input Schmitt NAND, were used, see Figure 2.16. These run at about 230 kHz, providing from the *MN3102*s a clock frequency for each BBD of around 115 kHz. The output waveform of the VCOs is distinctly asymmetrical, and varies with the LTP control input. But the *MN3102* device turns this into two antiphase non-overlapping clock waveforms with near unity mark/space ratios.

Differential delays

The LTP provided differential control, by subtracting a greater or lesser amount from the available charging current via the 27K

80 Analog circuits cookbook

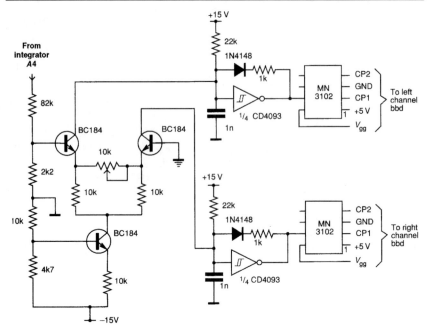

Figure 2.16 *Showing the differentially controlled VCOs driving the clock generators which service the BBD chips*

resistor, at the input of each VCO, such that as one VCO frequency increased the other reduced by the same percentage (at least, to a first approximation), see Figure 2.16. The BBD provides delays of 2.56 to 51.2 ms for clock frequencies in the range 200 kHz down to 10 kHz, so at the 115 kHz clock frequency used, the delay is nominally 4.45 ms. So to provide the required ±150 μs delay variation for a head movement of 45°, the frequency of the VCOs must be varied 0.15/4.45, or about ±3.4%. As this is but a small variation, the integrator output is attenuated before being applied to the LTP, the transconductance of which is adjustable by means of a 10K pot between the emitters. This pot provides an adjustment for the spacing between the ears of a listener, a fat-headed person will require a lower resistance setting of the pot than a narrow-minded type.

The non-overlapping clocks from each *MN3102* are applied to the corresponding *MN3207* BBD, which also each receive an audio input, see Figure 2.17(a). The delayed audio output from each BBD is applied to a three pole Chebychev filter, to suppress the clock ripple which appears in the BBD outputs. The filters are of a slightly unconventional kind, taking into account the output impedance of the BBDs, the input capacitance of the opamps, circuit strays, etc., so the capacitor values are not what you would obtain from the usual tables of normalised filters. Nevertheless, the response is flat to within 1 dB to beyond 15 kHz, 4 dB down at 20 kHz and already 33 dB down at 50 kHz.

The output filter opamps could not be expected to cope happily with the loads imposed by 32 Ω headphones, so a dual audio amplifier was used. This was the National Semiconductor *LM4880* Dual 250 mW Audio Power Amplifier, which operates on a single supply rail in the range 2.7–5.5 V. On a 5 V supply it provides 85 mW continuous average power into 32 Ω or 200 mW into 8 Ω, at 0.1%THD at 1 kHz. It features a shutdown mode which reduces the current drain from a typical 3.6 mA (no-signal quiescent) to around a microamp. For speed and convenience, the 'Boomer®' evaluation board, carrying the small outline version of the device was used, the circuit being as in Figure 2.17(b), the output coupling capacitors C_o being each two 100 μF electrolytics in parallel. Strapping the shutdown input to V_{DD} activates the shutdown feature, but as this was not required, the SD pad was strapped to ground.

Testing the kit

During design and construction, which proceeded in parallel, each section of circuitry was tested for functionality as it was added, starting with A_1 and working through to the audio output stage. But any serious overall evaluation of the scheme was obviously not possible until the whole equipment was complete. As mentioned earlier, the ac coupling at A_1 was discarded due to extended settling problems, the alternative dc coupling being adequate for an experimental set-up.

With the circuitry complete, a 250 Hz sinewave was applied to the two audio input channels strapped in parallel, the offset pot having been set up for zero output at A_3 when the gyro pointing was stationary, and the integrator output zeroed. Strapping the two inputs together provided a path for a little leakage of BBD clock frequency between devices, resulting in some low level 'birdies' being audible in the background, but these were ignored at this stage. On turning the head to either side, a most bizarre effect was noted. The pitch of the sound in the advancing ear (the right ear when turning the head to the left) momentarily rose whilst that in the other ear fell. At this point I realised that the attenuator between the integrator output and the LTP input had been omitted. The result was an enormous transient delay (phase) change in the signal, resulting in Doppler effect shifting of the frequency, as would indeed occur on turning the head if one's ears were a few tens of metres apart!

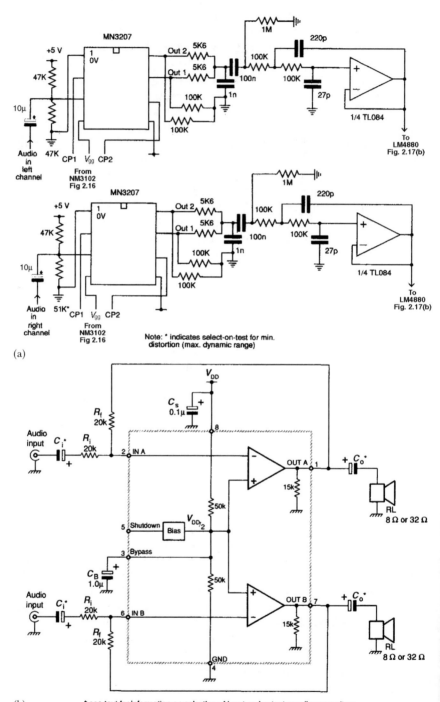

(a)

(b) * see text for information on selection of input and output coupling capacitors

With a suitable degree of attenuation added, as shown in Figure 2.16, the LTP emitter pot was adjusted to give +/–0.15 ms delay in one channel and –/+0.15 ms in the other for a 45° rotation of the head. The result was quite distinct; whilst facing front, the sound appeared to be arriving centrally, but from the right as the head was turned to the left and vice versa. Interestingly, the sound in the ear nearest the front actually sounded louder than that in the other ear, although of course the two signals were identical, except in phase. Evidently the ear/brain system is quite capable of resolving differential times of arrival of sound of the order of 100 µs.

Next, tests were carried out using programme material, from an FM radio. The signal was taken via a couple of 2 pin DIN speaker plugs from the set's external speaker outlets. Taking the signal from two separate low impedance outputs like this largely suppressed the birdies mentioned earlier. With reception switched to mono, programme material of all sorts behaved in exactly the same way as the continuous sinewave, the 'direction' of the source being readily identifiable. Much the same applied to speech in stereo, but since a microphone is usually used which is near (or actually on) the speaker, stereo speech is usually virtually mono anyway. However, disappointingly, results with an extended sound source, such as orchestral music in stereo, were not noticeably amenable to 'externalisation' by the system. The sound stage remained doggedly stuck to the head, turning with it. The reason for this is not clear to me, though some knowledgeable reader may well be able to provide enlightenment. Possibly the ear/brain system is so dominated by the abundance of positional information cues contained in a stereo signal, that it cannot but hear the sound as coming from a sound stage fixed relative to the head. Whatever the explanation, the scheme is virtually ineffective on stereo material. But that's engineering for you, the results of an experiment are what they are, not one might like them to be. Hypotheses have to fit the facts, not the other way round.

Figure 2.17 *(a) The BBD audio delay stages, followed by three pole Chebychev lowpass filters to remove clock ripple from the output of the BBDs. (b) The audio output stage, using an LM4880 Dual 250 mW audio power amplifier with shutdown mode (not used in this application). Note: If the sound stage moves to the left instead of the right when the head is turned to the left, the audio connections between (a) and (b) should be interchanged*

Some active filters

Active filters is a very wide subject. Whole books have been written about the topic. This short article looks at one or two of the common ones, and one or two of the less common. It concludes with details of the useful and economical second-order section known as the SAB – single active biquad – design equations for which can be found in Ref. 6.

Filter variations

Introduction

Many applications call for the filtering of signals, to pass those that are wanted, and to block those that are outside the desired passband. Sometimes digital filtering is appropriate, especially if the signals are in digital form already, but oftentimes, analog filters suffice – indeed are the only choice at RF. At lower frequencies, where inductors would be bulky, expensive and of low Q, active filters are the usual choice. Some of these are documented in every textbook, but there are some useful variations upon them which are less well known. This article explores one or two of these.

A basic active filter

Probably the best known active filter is the Sallen and Key second-order circuit, the lowpass version of which is shown in Figure 2.18. (Interchanging the Cs and Rs gives a highpass version.) There has been considerable discussion recently of its demerits, both in regard

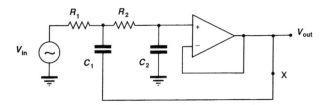

Figure 2.18 *The Sallen and Key second-order lowpass active filter. Cut-off ('corner') frequency is given by $f_o = 1/(2\pi C_1 C_2 R_1 R_2)$ and $Q = \frac{1}{2}\sqrt{(C_1/C_2)}$ and dissipation $D = 1/Q$. For a maximally flat amplitude (Butterworth) design, $D = 1.414$, so $C_1 = 2C_2$. The Butterworth design exhibits no peak, and is just 3 dB down at the corner frequency*

to noise and distortion, from Dr D. Ryder and others in the Letters section of *EW+WW*, see the November 1995 to April 1996 issues inclusive. But for many purposes it will prove adequate, having the minor advantage of very simple design equations. Moreover, the circuit is canonic – it uses just two resistors and two capacitors to provide its twopole response.

Being a second-order circuit, at very high frequencies the response falls away forever at 12 dB per octave – at least with an ideal opamp. (In practice, opamp output impedance rises at high frequencies, due to the fall in its open loop gain, resulting in the attenuation curve levelling out, or even reversing.) In the maximally flat amplitude response design, at frequencies above the cut-off frequency, the response approaches 12 dB/octave asymptotically, from below. At dc and well below the cut-off frequency, the response is flat, being 0 dB (unity gain), again a value the response approaches asymptotically from below. The corner formed by the crossing of these two asymptotes is often called, naturally enough, the 'corner frequency'. The corner or cut-off frequency f_0 is given by $f_0 = 1/(2\pi\sqrt{\{C_1 C_2 R_1 R_2\}})$ where, usually, $R_1 = R_2$.

The dissipation factor $D = 1/Q$ where $Q = \frac{1}{2}\sqrt{(C_1/C_2)}$ and for a maximally flat amplitude (Butterworth) design, $D = 1.414$, so $C_1 = 2C_2$. The Butterworth design exhibits no peak, and is just 3 dB down (i.e. $V_{out}/V_{in} = 0.707$, or equal to Q) at the corner frequency. If $C_1 > 2C_2$, then there is a passband peak in the response below the corner frequency, being more pronounced and moving nearer the corner frequency as the ratio is made larger. This permits the design of filters with four or six poles, or of even higher order, consisting of several such stages, all with the same corner frequency but each with the appropriate value of Q.

It is easy to see that the low frequency gain is unity, by simply removing the capacitors from Figure 2.18, for at very low frequencies their reactance becomes so high compared to R_1, R_2, that they might as well simply not be there. At a very high frequency, way beyond cut-off, C_2 acts as a near short at the non-inverting (NI) input of the opamp, resulting in the lower plate of C_1 being held almost at ground. As C_1 is usually greater than C_2, it acts in conjunction with R_1 as a passive lowpass circuit well into its stopband, resulting in even further attenuation of the input. At twice this frequency, both of these mechanisms will result in a halving of the signal, which thus falls to a quarter of the previous value, i.e. the roll-off rate is 20 $\log(1/4)$ or –12 dB/octave. But what about that peak in the passband, assuming there is one?

The best way to approach this is to break the loop at point X (in Figure 2.18) and consider what happens to a signal V'_{in}, going round the loop, having removed the original V_{in}. Note that as the source in Figure 2.18 is assumed to have zero internal resistance, it has been replaced by a short circuit in Figure 2.19. To V'_{in}, C_1 with R_1 now forms a passive lead circuit – highpass or bass cut. The resultant voltage across R_1 is applied to C_2, R_2, a passive lag circuit – lowpass or top cut. Each of these responses exhibits a 6 dB/octave roll-off in the

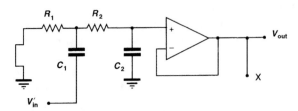

Figure 2.19 *Breaking the loop and opening it out helps to understand the circuit action (see text)*

stopband, as shown in Figure 2.20. Thus the voltage reaching the NI input of the opamp at any frequency will be given roughly by the sum of the attenuation of each CR section (actually rather more, as C_2R_2 loads the output of the C_1R_1 section), as indicated by the dotted line in Figure 2.20. At the frequency where the highpass and lowpass curves cross, the attenuation is a minimum, and the phase shift is zero since the lag of one section cancels the lead of the other.

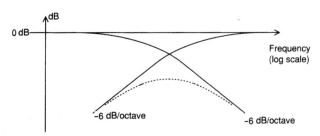

Figure 2.20 *Cascaded lowpass and highpass CR responses, and their resultant, (dotted)*

If now C_1 is made very large, the bass cut will only appear at very low frequencies – the highpass curve in Figure 2.20 will shift bodily to the left. If in addition C_1 is made very small, the top cut will appear only at very high frequencies – the lowpass curve will shift bodily to the right. Thus the curves will cross while each still contributes very little attenuation, so the peak of the dotted curve will not be much below 0 dB, unity gain. Consequently, at this frequency the voltage at

X is almost as large as V'_{in}, and in phase with it. The circuit can almost supply its own input, and if disturbed in any way will respond by ringing at the frequency of the dotted peak, where the loop phase shift is zero.

But however large the ratio C_1/C_2, there must always be some attenuation, however small, between V'_{in} and the opamp's NI input, so the circuit cannot oscillate, although it can exhibit a large peak in its response, around the corner frequency. In fact, if the peak is large enough, the filter's response above the corner frequency will approach the −12 dB/octave asymptote from above, and below the corner frequency will likewise approach the flat 0 dB asymptote from above.

Variations on a theme

The cut-off rate can be increased from 12 dB/octave to 18 dB/octave by the addition of just two components; a series R and a shunt C to ground between V_{in} and R_1. And such a third-order section can be cascaded with other second-order section(s) to make filters with five, seven, nine poles, etc. Normalised capacitor values for filters from two to ten poles for various response types (Butterworth, Chebychev with various passband ripple depths, Bessel, etc.) have been published in Refs 1 and 2, and in many other publications as well. However, these tables assume $R_1 = R_2$ (=the extra series resistor in a third-order section), with the Q being set by the ratio of the capacitor values. This results in a requirement for non-standard values of capacitor, which is expensive if they are specially procured, or inconvenient if made up by paralleling smaller values.

Whilst equal value resistors are optimum, minor variations can be accommodated without difficulty, and this can ease the capacitor requirements. Ref. 3 gives tables for the three resistors and three capacitors used in a third-order section, with the capacitors selected from the standard E3 series (1.0, 2.2, 4.7) and the resistors from the E24 series, for both Butterworth and Bessel (maximally flat delay) designs.

The Kundert filter

The formula for the Q of the Sallen and Key filter is $Q = \frac{1}{2}\sqrt{(C_1/C_2)}$, so given the square root sign and the $\frac{1}{2}$ as well, one finishes up with rather extreme ratios of C_1 to C_2, if a high Q is needed, as it will be in a high-order Chebychev filter. In this case, the Kundert circuit of Figure 2.21 may provide the answer. The additional opamp buffers the second CR from the first, so that the attenuation at any frequency represented by the dotted curve in Figure 2.20 is now exactly equal to

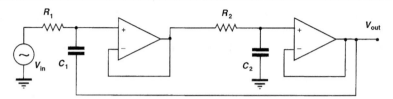

Figure 2.21 *The Kundert filter, a variant of the Sallen and Key, has some advantages*

the sum of the other two curves. Removing the loading of C_2R_2 from C_1R_1 removes the $1/2$ from the formula, which is now $Q = \sqrt{(C_1/C_2)}$ – assuming $R_1 = R_2$. And due to the square root sign, the required ratio of C_1 to C_2 for any desired value of Q is reduced by a factor of four compared to the Sallen and Key version.

A further advantage of this circuit is the complete freedom of choice of components. Instead of making $R_1 = R_2$ and setting the Q by the ratio of C_1 to C_2, the capacitors may be made equal and the Q set by the ratio of R_1 to R_2, or both Cs and Rs may differ, the Q being set by the ratio of C_1R_1 to C_2R_2. Given that dual opamps are available in the same 8 pin DIL package as single opamps, the Kundert version of the Sallen and Key filter, with its greater freedom of choice of component values, can come in very handy for the highest Q stage in a high-order filter.

The equal C filter

In addition to filtering to remove components outside the wanted passband, signals also frequently need amplification. The basic Sallen and Key circuit only provides unity gain, and with this arrangement equal resistors are optimum. (For, due to the loading of the second stage on the first, if R_2 is increased to reduce the loading, then C_2 will have to be even smaller, while if R_2 is decreased to permit a larger value of C_2, the loading on C_1R_1 increases.)

Where additional signal amplification is needed, there is no reason why some of this should not be provided within a filtering stage and Figure 2.22 shows such a circuit. Clearly the dc and low frequency gain is given by $(R_A + R_B)/R_B$. A convenience of this circuit is that the ratio R_A to R_B can be chosen to give whatever gain is required (within reason), with C_1, C_2, R_1, R_2, chosen to give the required corner frequency and Q. An analysis of this most general form of the circuit can be found in Ref. 4. If there were a buffer stage between R_1 and R_2 as in Figure 2.21, and the two CR products were equal, then at a frequency of $1/(2\pi CR)$ there would be exactly 3 dB attenuation round the loop due to each CR. So if R_A were to equal R_B, giving 6 dB gain

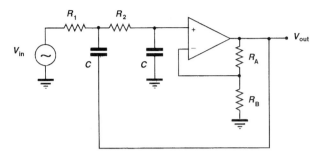

Figure 2.22 *The equal C version of the Sallen and Key circuit*

in the opamp stage, there would be no net attenuation round the loop and the Q would equal infinity – you have an oscillator. Without the buffer opamp, the sums are a little more complicated due to the second CR loading the first. But the sums have all been done, and the normalised values for R_1 and R_2 (values in ohms for a cut-off frequency of $1/2\pi$ Hz, assuming $C = 1$ F) are given in Ref. 5 for filters of one to nine poles, in Butterworth, Bessel and 0.1 dB-, 0.5 dB- and 1 dB-Chebychev designs. (For odd numbers of poles, this reference includes an opamp buffered single pole passive CR, rather than a 3 pole version of the Sallen and Key filter, as one of the stages.) To convert to a cut-off frequency of, say, 1 kHz, divide the resistor values by 2000π. Now, to obtain more practical component values, regard the ohms figures as MΩ and the capacitors as 1 µF. As the values are still not convenient, scale the capacitors in a given section down by, say, 100 (or any other convenient value), and the resistors up by the same factor.

Ref. 5 also gives the noise bandwidth of each filter type. The noise bandwidth of a given filter is the bandwidth of a fictional ideal brick wall-sided filter which, fed with wideband white noise, passes as much noise power as the given filter. Ref. 5 also gives, for the Chebychev types, the 3 dB bandwidth. Note that for a Chebychev filter, this is not the same as the specified bandwidth (unless the ripple depth is itself 3 dB). For a Chebychev filter the bandwidth quoted is the ripple bandwidth; e.g. for a 0.5 dB ripple lowpass filter, the bandwidth is the highest frequency at which the attenuation is 0.5 dB, beyond which it descends into the stopband, passing through –3 dB at a somewhat higher frequency.

Other variants

In the Sallen and Key filter, the signal appears at both inputs of the opamp. There is thus a common mode component at the input, and this can lead to distortion, due to 'common mode failure', which, though small, may be unacceptable in critical applications. Also, as already mentioned, the ultimate attenuation in the stopband will often be limited by another non-ideal aspect of practical opamps – rising output impedance at high frequencies, due to the reduced gain within the local NFB loop back to the opamp's inverting terminal. Both of these possibilities are avoided by a different circuit configuration, shown (in its lowpass form) in Figure 2.23(a). This is variously known as the infinite gain multiple feedback filter, or the Rausch filter, and it has the opamp's NI terminal firmly anchored to ground – good for avoiding common mode failure distortion. Another plus point is that at very high frequencies, C_1 short circuits the signal to ground, whilst C_2 shorts the opamp's output to its inverting input – good for maintaining high attenuation at the very highest frequencies. The design equations and tabulated component values are available in published sources; the filter is well known and is shown here just as a stepping stone to a less well-known second-order filter section. This is the SAB (single active biquad) lowpass section, which possesses a finite zero in the stopband.

In some filtering applications, the main requirement is for a very fast rate of cut-off, the resultant wild variations in group delay not being important. The Chebychev design provides a faster cut-off than the Butterworth, the more so, the greater ripple depth that can be tolerated in the passband. But the attenuation curve is monotonic, it just keeps on going down at (6n) dB/octave, where n is the order of

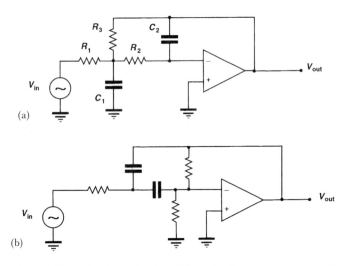

Figure 2.23 (a) The mixed feedback or 'Rausch' filter – lowpass version. (b) The mixed feedback or 'Rausch' filter – bandpass version

the filter (the number of poles), not reaching infinite attenuation until infinite frequency. A faster cut-off still can be achieved by a filter incorporating one or more finite zeros, frequencies in the stopband at which the response exhibits infinite attenuation – a notch. In a design with several such notches, they can be strategically placed so that the attenuation curve bulges back up in between them to the same height each time. Such a filter, with equal depth ripples in the passband (like a Chebychev) but additionally with equal returns between notches in the stopband is known as an 'elliptic' or 'Caur' filter.

In a multipole elliptic filter, each second-order section is designed to provide a notch, but beyond the notch the attenuation returns to a steady finite value, maintained up to infinite frequency. The nearer the notch to the cut-off frequency, the higher the level to which the attenuation will eventually return above the notch frequency. So for the highest cut-off rates, whilst still maintaining a large attenuation beyond the first notch, a large number of poles is necessary. It is common practice to include a single pole (e.g. an opamp buffered passive CR lag) to ensure that, beyond the highest frequency notch, the response dies away to infinite attenuation at infinite frequency, albeit at a leisurely –6 dB/octave.

The elliptic filter

The building blocks for an elliptic lowpass filter consist of second-order lowpass sections of varying Q, each exhibiting a notch at an appropriate frequency above the cut-off frequency.

A number of designs for such a section have appeared, based on the TWIN TEE circuit, but they are complex, using many components, and hence difficult to adjust. An alternative is provided by the SAB section mentioned earlier. This can be approached via the Rausch bandpass filter, which can be seen (Figure 2.23(b)) to be a variant on the Rausch lowpass design of Figure 2.23(a). Clearly, due to the capacitive coupling, the circuit has infinite attenuation at 0 Hz, and at infinite frequency, the capacitors effectively short the opamp's inverting input to its output, setting the gain to zero. Either side of the peak response, the gain falls off at 6 dB per octave, the centre frequency Q being set by the component values. If the Q is high, the centre frequency gain will be well in excess of unity.

Figure 2.24 shows the same circuit with three extra resistors (R_2, R_3 and R_6) added. Note that an attenuated version of the input signal is now fed to the NI input of the opamp via R_2, R_3. Consequently, the circuit will now provide finite gain down to 0 Hz; it has been converted into a lowpass section, although if the Q is high there will

92 Analog circuits cookbook

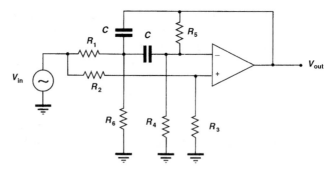

Figure 2.24 *The SAB circuit, with finite zero (or notch, above the passband)*

still be a gain peak. If the ratio of R_5 to R_4 is made the same as R_2 to R_3, then the gain of the opamp is set to the same as the attenuation suffered by the signal at its NI terminal, so the overall 0 Hz gain is unity. If the other components are correctly chosen, the peak will still be there, but at some higher frequency, the signal at the opamp's inverting input will be identical in phase and amplitude to that at the NI input. The components thus form a bridge which is balanced at that frequency, resulting in zero output from the opamp, i.e. a notch.

Figure 2.25 shows a 5 pole elliptic filter using SAB sections, with a 0.28 dB passband ripple, a −3 dB point at about 3 kHz and an attenuation of 54 dB at 4.5 kHz and above. The design equations for elliptic filters using SAB sections are given in Ref. 6. The design equations make use of the tabulated values of normalised pole and zero values given in Ref. 7.

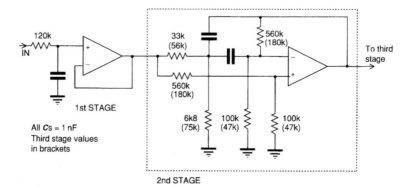

Figure 2.25 *A 5 pole elliptic filter with 0.28 dB passband ripple and an attenuation of 54 dB at 1.65 times the cut-off frequency and upwards. The −3 dB point is 3 kHz, approx. All capacitors C = 1 nF, simply scale C for other cut-off frequencies*

Some other filter types

Simple notch filters – where the gain is unity everywhere either side of the notch – can be very useful, e.g. for suppressing 50 Hz or 60 Hz hum in measurement systems. The passive TWIN TEE notch is well known, and can be sharpened up in an active circuit so that the gain is constant, say, below 45 Hz and above 55 Hz. However, it is inconvenient for tuning, due to the use of no less than six components. An ingenious alternative (Ref. 8) provides a design with limited notch depth, but compensating advantages. A notch depth of 20 dB is easily achieved, and the filter can be fine tuned by means of a single pot. The frequency adjustment is independent of attenuation and bandwidth.

Finally, a word on linear phase (constant group delay) filters. These are easily implemented in digital form, FIR filters being inherently linear phase. But most analog filter types, including Butterworth, Chebychev and elliptic, are anything but linear phase. Consequently, when passing pulse waveforms, considerable ringing is experienced on the edges, especially with high-order filters, even of the Butterworth variety. The linear phase Bessel design can be used, but this gives only a very gradual transition from pass- to stopband, even for quite high orders. However, a fact that is not widely known is that it is possible to design true linear phase filters in analog technology, both bandpass (Ref. 9) and lowpass (Ref. 10). These can use passive components, or – as in Ref. 10 – active circuitry.

References

1. Shepard, R.R. (1969) Active filters: part 12 – Short cuts to network design. *Electronics*, Aug. 18, pp 82–91.
2. Williams, A.B. (1975) *Active Filter Design*. Artech House Inc.
3. *Linear Technology Magazine*, May 1995, p. 32.
4. Huelsman, L.P. (1968) *Theory and Design of Active RC Circuits*. McGraw-Hill Book Company, p. 72.
5. Delagrange, A. (1983) Gain of two simplifies LP filter design. *EDN*, 17 March, pp. 224–228. (Reproduced in *Electronic Circuits, Systems and Standards*, Ed. Hickman, Butterworth-Heinemann 1991, ISBN 0 7506 0068 3.)
6. Hickman, I. (1993) *Analog Electronics*. Butterworth-Heinemann, ISBN 0 7506 1634 2.
7. Zverev, A.I. (1967) *Handbook of Filter Synthesis*. John Wiley and Sons Inc.
8. Irvine, D. (1985) Notch filter. *Electronic Product Design*, May, p. 39.

9. Lerner, R.M. (1964) Bandpass filters with linear phase. *Proc. IEEE*, March, pp. 249–268.
10. Delagrange, A. (1979) Bring Lerner filters up to date: replace passive components with opamps. *Electronic Design* **4**, 15 February, pp. 94–98.

> **Video dubbing box**
>
> As every camcorder owner knows, a car engine sound track does nothing to enhance scenery filmed from a moving vehicle. Ian Hickman's mixing circuit is designed not only to remove unwanted sound but also to replace it with whatever you want.

Camcorder dubber

On a winter holiday recently, we took a camcorder with us for the first time. Not wanting to tie up the expensive HI8 metal tapes for ever, and play holiday movies through the camcorder, the obvious step was to transfer the material to VHS tapes. But that raised the question of what to do about the sound.

In earlier times, with 8 mm home movies, usually there was no sound, or at least it was added afterwards, by 'striping' the film. The camcorder, by contrast, gives one stereo sound, whether you want it or not. This is fine when shooting a street carnival, or for catching everything but the smell of a loco in steam. But often – as when shooting through the windscreen of a car in the snow covered Troodos mountains – the sound is more of a nuisance than a help. So we had taken the step of bringing back with us a tape of Greek instrumental Sirtaki and Bouzouki music, for use as background music.

A simple appliqué box

All that was needed now was a gadget to permit the sound on the VHS tape to originate from either the camcorder, or from a cassette recorder, at will.

Further consideration made it clear that it should be possible to 'cross-fade' the two sound sources, to avoid any clicks due to switching transients. And a further useful facility would be a microphone input, so that comments, or at least a simple introduction, could be added to the soundtrack. To avoid further switching arrangements, the microphone input should operate a 'ducking circuit', to reduce the volume of the background music when speaking. The necessary circuitry was soon sketched out and built up.

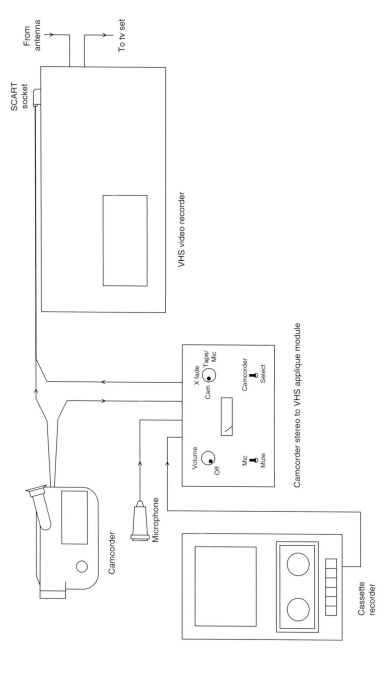

Figure 2.26 *Showing the various pieces of kit interconnected. The sound from the camcorder is routed via the appliqué box. The latter permits cross-fading of the camcorder sound output with sound from tape and/or microphone*

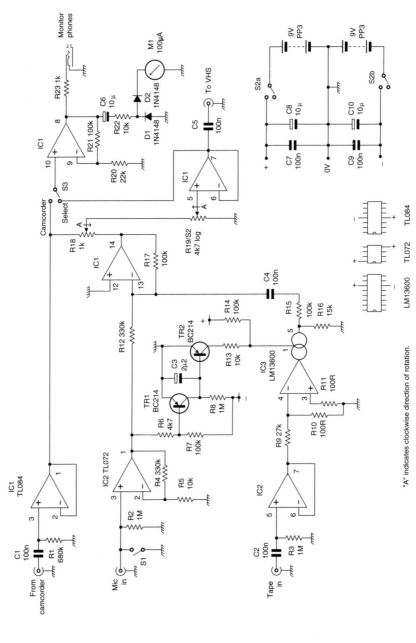

Figure 2.27 *The circuit diagram of the appliqué box, for HI8 to VHS sound dubbing*

At the end of the day, the result was a suitable appliqué box, interconnecting the various items of kit as illustrated in Figure 2.26.

The circuitry

Figure 2.27 shows the circuit of the appliqué box. Unity gain buffers are provided for the camcorder sound and the input from the cassette recorder's DIN connector (which turned out to be at much the same level), whilst the microphone input buffer also provides 30 dB of gain. The microphone and tape inputs are summed at the virtual earth, pin 13, of IC_1, and applied to one end of a 1K linear potentiometer R_{18}. The buffered 'HI8' sound from the camcorder is connected to the other end, the wiper of R_{18} being connected to volume control R_{19}, a 4k7 log pot. Thus the sound output from buffer IC_1 pin 7 can be cross-faded at will between the HI8 and tape/microphone inputs, and its level adjusted from normal down to zero. The audio-in phono plug end of the camcorder's SCART-to-video recorder lead, normally connected straight to the camcorder's sound output, is connected to SK_4.

The buffered audio from tape is fed via one half of an *LM13600* dual transconductance amplifier, IC_3. The *LM13700* is often preferred for audio work. This is because that device's output Darlington buffers exhibit no shift in dc level with change of transconductance. By contrast, as the transconductance is increased or decreased, the *LM13600*'s output buffers are biased up to a greater or lesser degree, in sympathy. The arrangement results in a faster slew rate, handy in circuits where fast settling is needed. But it can result in 'pops' in an audio circuit, when there are rapid changes in gain. However, in this application the Darlington output buffers are not used, so either device will do.

The transconductance of IC_3, and hence its gain, is set by the bias current I_{ABC} injected into pin 1. This consists of two components, the larger proportion coming via R_{13} (TR_2 is normally bottomed), with about another 25% or so coming from the positive rail, via R_{14}. When a voice-over output from the mike appears, the negative-going peaks at pin 1 of IC_2 bottom TR_1. This discharges C_3 and removes the base current from TR_2. The gain of IC_3 thus drops by some 12 dB or more, this proving a suitable degree of ducking.

The microphone used was a small dynamic type with a 50K output impedance. In fact, it needed only 20 dB of gain to raise its output to the same level as that from the camcorder and cassette. The extra gain, together with a little forward bias for TR_1 via R_7, was incorporated to provide reliable operation of the ducking function. The extra microphone circuit gain was simply disposed of by making R_{12} 330K, as against 100K at R_{15} and R_{17}.

S_1 shorts the mike when not needed, preventing adventitious extraneous noises appearing on the soundtrack of the dubbed tape. The output of the fourth section of IC_1, at pin 8, is used as a buffer to drive monitor phones, which can be plugged into SK_5. It also drives a simple level monitor indicator, M_1. S_3 draws the output monitor/meter buffer's input either from the HI8 input (regardless of the settings of R_{18} and R_{19}), or from the current 'Select' input, be it HI8, cassette or microphone.

The whole circuit was mounted in a 'recycled' metal case, i.e. one resurrected from a redundant earlier project. Power is supplied by two internal PP3 9 V layer type batteries, the ON/OFF switch being ganged with R_{19}. C_7 and C_9 were in fact duplicated adjacent to each IC, in accordance with good practice.

Using the appliqué box

When transferring video from HI8 to a VHS tape in the video recorder, the latter is set to use the SCART socket as the programme source. On our video recorder, this is achieved by setting its channel number to 0, which brings up the legend 'AV' in the channel number display – a fairly standard arrangement, I imagine.

When viewing video tapes, our TV is usually supplied with baseband video via a SCART interconnection, avoiding further loss of picture quality by transfer at RF on channel 36. The SCART socket is not available when dubbing, as it is required for the lead from the camcorder and appliqué box. But setting the TV to the channel number tuned to channel 36 enables visual monitoring of the HI8 output as it is recorded, and also of the 'Select' sound via the dubbing box. Thus the main use for the phone monitor is to keep an ear on the original HI8 soundtrack, ready to cross-fade to it when appropriate.

The circuit shown in Figure 2.27 is for my particular collection of kit. Depending on the particular microphone and cassette recorder (or CD player – or even record deck) used, different gain settings of the input buffers may be required. Here, the level meter M_1 is handy, as indicating the typical level of HI8 sound out of the camcorder. Stereo enthusiasts with a suitable video recorder can double up on IC_1 and use a quad opamp in place of a dual at IC_2, to provide stereo working. A second transconductance amplifier is of course already available in IC_3. However, stereo working for the voice-over channel would seem a little over the top.

Instead of cross-fading the two sound sources, R_{19} may alternatively be used in conjunction with R_{18} to fade one out and then the other in, if preferred. If voice-overs are going to be fairly infrequent, S_1 can usefully be a biased toggle, so that the microphone input is permanently muted, except when required.

3 Measurements (audio and video)

> ### *Ingenious video opamp*
>
> In an instrumentation amplifier, both inputs are high impedance and floating with respect to ground, but performance is limited to the sub-rf range. The opamp described here avoids that limitation, operating up to many tens of MHz.

Four opamp inputs are better than two

The INGENIous enginEER (in continental Europe the word for engineer is ingenieur) is always looking for elegant and economical solutions to design problems. Back in the 1970s when the RCA *CA3130* BiMOS opamp became available, it was clearly the answer to many an engineer's prayer, with its very high input impedance compared with the existing bipolar types.

I decided it was just the thing for the detector in a bridge circuit, but there was a snag. A bridge detector needs not only a high differential input impedance, but also both inputs must present a high

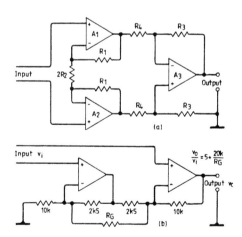

Figure 3.1 *Instrumentation amplifiers, floating high-impedance inputs. Circuits using (a) three opamps or (b) two opamps*

impedance to ground, to simulate the conventional floating detector circuit. With gain defining resistors fitted, this is no longer the case, but the amplifier cannot be used without them, since the open-loop gain times the offset voltage could result in the output being driven to one of the rails. Of course one could have a high impedance for both inputs with the usual instrumentation amplifier set-up of Figure 3.1(a), but why use three opamps if you can get away with fewer? The circuit of Figure 3.1(b) uses only two, but I was not aware of that particular circuit arrangement at the time. So I came up with the circuit of Figure 3.2, where an NFB loop around the amplifier is closed via one of the offset null terminals, leaving both the I and NI (inverting and non-inverting) input terminals free to float. With the offset null trimmed out, the circuit made a fine detector for a dc excited resistance bridge, the *CA3130*'s 90 dB typical CMRR (common mode rejection ratio) resulting in negligible error with change in bridge ratios. But it also made a fine inductance bridge, the values in Figure 3.2 giving a 100 μH full scale range. The 100 Ω standard resistor R_s was switchable to 1 kΩ or 10 kΩ giving 1 and 10 mH ranges, and then switching C_s to 100 nF gave 0.1, 1 and 10 H ranges. The opamp's input stage is outside the NFB loop, so its gain will vary somewhat with temperature, but for a bridge detector that is not important; in any case a wide range of gain control was needed to cope with the different bridge ratios and this was supplied by the 100 kΩ log sensitivity pot. The CMRR of the *CA3130* at 1592 Hz (ω = 10^4 rad/sec) is not stated in the data but seemed adequate for the purpose, and the resultant simple *RCL* bridge served me well for many years.

Recently the *LT1193* and *LT1194* video difference amplifiers caught my eye in *Linear Technology* (1991) and I received samples of

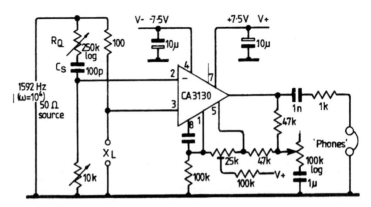

Figure 3.2 *Inductance bridge with a 50 Ω source providing a dc path to ground*

them from the manufacturer, Linear Technology Corporation. They are part of the *LT119x* family of low-cost high-speed fast-settling opamps, which includes devices with gain-bandwidth products up to 350 MHz, all with a 450 V/μs slew rate. With this sort of performance, you won't be surprised to learn that the parts use bipolar technology.

The *LT1193* and *LT1194* video difference amplifiers differ from the other members of the family in that they have two pairs of differential input terminals, so that the gain-defining NFB loop can be closed around one pair, leaving the other pair floating free. The input impedance of the *LT1193* is typically 100 kΩ in parallel with 2 pF at either the I or the NI input. Figure 3.3(a) shows the device used as an 80 MHz (−3 dB) bridging amplifier, tapped across a 75 Ω coaxial video distribution system. This arrangement is clearly much more economical than the usual alternative of terminating the incoming signal in a video repeater amplifier housed in a distribution box, and providing a fan-out of several outputs, for local use and for the on-going run to the next distribution box. However, although the signal in the cable is nominally unbalanced (i.e. ground referenced), in practice there are ground loops between pieces of equipment, and high frequency common mode noise is often induced in the cable. So the bridging amplifier at each tap location requires a high CMRR at high frequency.

Figure 3.3(b) shows a 5 MHz signal recovered from an input with severe common mode noise, illustrating that the CMRR is maintained at high frequencies. Whereas my floating input *CA3130* circuit's gain was not well defined, the input stage being outside the gain defining NFB loop, the *LT1193* does not suffer from this disadvantage. Its two input stages are provided with identical emitter-to-emitter degeneration resistors (Figure 3.3(c)), so that the gain at the I and NI inputs (pins 2 and 3) is the same as that defined at the reference and feedback inputs, pins 1 and 8. The gain error is typically 0.1% while the differential gain and phase errors at 3.58 MHz are 0.2% and 0.08° peak to peak respectively. While excellent as double-terminated 75 Ω cable drivers, the *LT1193/4* are capable of stably driving 30 pF or more of load capacitance with minimal ringing.

The *LT1193* features a unique facility, accessed by pin 5, that enables the amplifier to be shut down to conserve power, or to multiplex several amplifiers onto a single cable. Pin 5 is left open-circuit for normal operation, but pulling it to the negative supply rail gates the output off within 200 ns leaving the output tri-stated and typically reducing the dissipation from 350 mW (with +5 V and −5 V rails) to 15 mW. The *LT1194* (whose gain is internally set at ×10) has a different party trick, made possible by bringing out the emitters of

102 Analog circuits cookbook

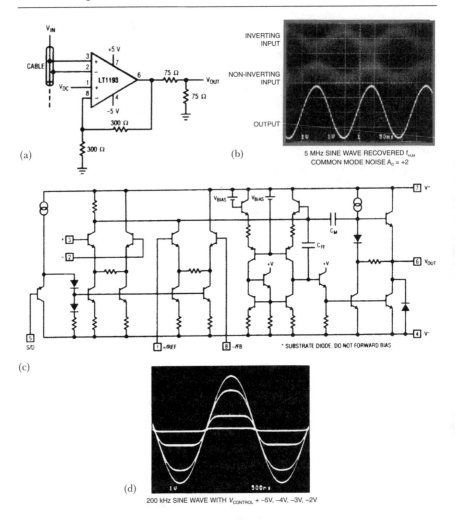

Figure 3.3 (a) Cable sense amplifier for loop through connections with dc adjust. (b) Recovered signal from common mode noise. (c) LT1193 simplified schematic. (d) Sinewave reduced by limiting the LT1194

the input stage's constant current tail transistors. This enables the input stage's current to be reduced by degrees, limiting the available output swing (Figure 3.3(d)). This technique allows extremely fast limiting action.

The applicability of the fully floating input stage of the *LT1193* to my old bridge circuits was immediately apparent, and on seeing that the device's CMRR was still in excess of 55 dB at 1.592 MHz (Figure 3.4(a)), it was clear that the bridge could be run at $\omega = 10^7$, enabling much lower values of inductance to be measured. So the circuit of Figure 3.4(b) was hastily built and tested. With the values shown, inductances up to 200 nH can be measured, and the circuit was tried out using a 'Coilcraft' (see Ref. 1) five and a half turn air-cored inductor of 154 nH, type 144-05J12 (less slug). I have not yet succeeded in finding a non-inductive 20 Ω potentiometer for R_v, so balance was achieved by selecting resistors on a trial and error basis. The bridge balanced with R_v equal to 15 Ω in parallel with 220 Ω, and with a 180 pF capacitor as the tan δ 'control'. These values give the inductance as 145 nH and the Q at 1.592 MHz as 5.5. The measured value of inductance is a little adrift, but that is not surprising, given the bird's nest construction. Indeed, a quick check by connecting both inputs to the same side of the bridge showed that I was only getting 47 dB CMRR, even after removing the 100 nF capacitor decoupling the negative rail. This should have made things worse, not better. But then one must expect such oddities when using experimenter's plug board construction.

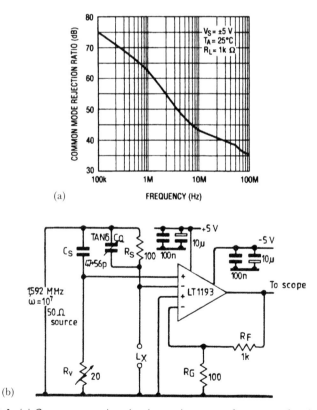

Figure 3.4 (a) Common mode rejection ratio versus frequency for the LT1193. (b) The 'hastily constructed' circuit using the LT1193 in a bridge application

The manufacturer's figure for the Q is 154 minimum at 40 MHz. If we assume that Q is proportional to frequency, the 'measured' Q is 138. But the manufacturer's figure of 154 is with the slug fitted, at mid-range, giving an inductance of 207 nH, so at 154 nH without the slug, a lower value of Q is only to be expected. In fact, the results from the bird's nest test bed are so encouraging that the circuit will now be rebuilt – properly!

References

1. Coilcraft, 1102 Silver Lake Rd, Cary, IL 60013 USA (312) 639-6400.
    Also in the UK at 21 Napier Place, Wardpark North, Cumbernauld, Glasgow G68 0LL.
2. *Linear Technology*, **1**, (2), October 1991.

Anti-alias filtering

Before applying an analog signal to an A-to-D converter, it is necessary to lowpass filter it to remove any components above half the sample rate – otherwise these may alias down into the bandwidth of interest. If the latter extends down to dc, then the filter must introduce no offset. This section describes a filter that fits the bill exactly.

DC accurate filter plays anti-alias role

Much signal processing nowadays, especially at audio and video frequencies, is carried out in DSP, a variety of digital signal processing chips providing a wide choice of speed, number of bits and architectures. But before a signal can be processed with a DSP it must be digitised, and before it is digitised it is advisable to lowpass filter it. Of course, the application may be such that no frequency components are expected in the signal at or above the Nyquist rate, but there is always the possibility of extraneous interference entering the system and thus it is a confident or more likely a foolhardy engineer who will dispense with an anti-alias filter altogether.

Many years ago such a filter might well have been passive *LC*, but these were advantageously displaced by active *RC* filters, which could do exactly the same job (subject to dynamic range limitations) much more cheaply. But like the *LC* filters, they were not easily variable or programmable. This practical difficulty was overcome by the arrival of the switched capacitor filter, although aliasing was now a possibility due to the time-discrete nature of the *SC* filter. However, as the clock frequency is fifty or a hundred times the filter's lowpass cut-off frequency, a simple single pole *RC* roll-off ahead of the filter often suffices, with another after to suppress clock frequency hash in the output. Even if a variable clock frequency is being used to provide a programmable cut-off, a fixed *RC* may still be enough if the range of cut-off frequency variation is only an octave or two, particularly if the following A-to-D converter uses only eight bits.

Aliasing problems are avoided entirely if a time continuous filter is used, and such filters are available in *IC* form requiring no external capacitors, for example the 8th order/4th order *MAX274/275* devices. The cut-off frequency and response type (Butterworth, Bessel, Chebychev, etc.) are programmed by means of external resistors. Although cut-off frequencies down to 1 kHz or lower are realisable with manageable resistor values, the cut-off frequency cannot be varied once set, though a limited choice of corner frequencies could (rather cumbersomely) be accommodated by selecting different sets of resistors by means of analog switches.

Good compromise

An interesting alternative filter type represents a sort of halfway house between pure time continuous filters and clock-tunable filters. The 'dc accurate' *MAX280* plus a few passive components makes a five pole lowpass filter with a choice of approximations to Butterworth,

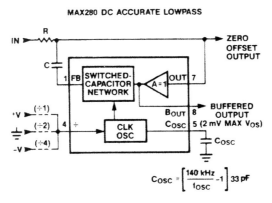

Figure 3.5 *Connecting up the MAX280 to act as a capacitance multiplier, with C appearing ever greater with progress up the stopband*

Bessel or Elliptic characteristics, and since the *RC* passive single pole is located right at the filter's input, it does duty as the anti-alias filter – providing 43 dB of attenuation at the Nyquist frequency. Figure 3.5 is a block diagram of this unusual filter arrangement, from which it can be seen immediately that the 'earthy' end of the *RC*'s capacitor goes not to ground but to a pin labelled FB (feedback). If it were grounded, the stopband response would show the usual 6 dB per octave roll-off, but in fact the chip acts as a capacitance multiplier, making *C* appear even greater as one moves higher up the stopband. The result is a fifth-order 30 dB/octave roll-off. Exactly how it works even the Maxim Applications Engineer was not entirely clear, but as he put it 'a lot of gymnastics goes on between pins 7 and 1'.

The filter's cut-off frequency is set by the clock frequency; this comes from a free-running internal oscillator which may alternatively be overridden by an external clock applied to the C_{osc} terminal, pin 5 (11) on the 8 (16) pin DIP package, and swinging close to the V+ and V– rails. Using no additional C_{osc}, the internal clock runs at 140 kHz nominal and as this can vary by as much as ±25% over the full range of supply voltages, it is as well to stabilise them. To check the clock frequency with a scope, turn the sensitivity up to maximum and just hold the probe near to the C_{osc} pin – even the 11 pF or so of a ×10 probe can pull the frequency down 20% if actually connected directly. With no additional C_{osc}, the filter's –3 dB point will be a little over 1 kHz with the divider ratio pin connected to V+. Alternatively, it may be connected to ground or V–, dividing the internal clock F_{osc} by two or four, lowering the cut-off frequency by one or two octaves. An external C_{osc} can be added if an even lower filter cut-off frequency is

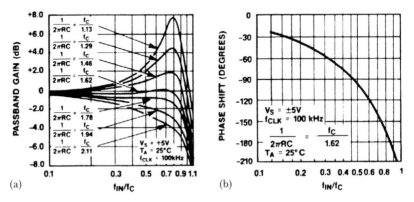

Figure 3.6 *Typical operating characteristics: (a) passband gains versus input frequency; (b) phase shift showing that this characteristic has already reached 180° at 0.85 of the 3 dB cut-off frequency* f_c

required. For a cut-off frequency higher than 1 kHz, an external clock of up to 4 MHz may be used. The filter's response shape in the region of the passband/stopband transition is determined by the relation between the clock frequency applied to the SC network and the time constant of the passive RC (Figure 3.6), which also shows the passband phase response. Note that, being a fifth-order network, the phase shift has already reached 180° at about 0.85 of the filter's 3 dB cut-off frequency f_c. Thus a notch filter is readily implemented using the circuit of Figure 3.7 (which corrects a misprint on the data sheet). I tried it out with $R = 39$ kΩ, $C = 6n2$, and $R_1 - R_4$ all 100 kΩ and obtained a nice deep notch at 890 Hz. On checking with a 'scope (that's when I found out about probe loading altering the clock frequency) the internal clock was found to be running at 105 kHz, as expected. Well below the notch frequency the gain of the circuit is ×2; well above – where the path through the filter is dead – it is ×1.

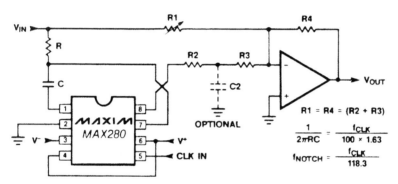

Figure 3.7 *The MAX280/LTC1062 used to create a notch. The input signal can be summed with the filter's output to create the notch*

Targeting fastest cut-off

As already noted, when used as a lowpass filter the response type is set by the CR time constant relative to the clock frequency, giving approximations to a Butterworth or Bessel response (Figure 3.8). However, where the fastest possible rate of cut-off is required in the stopband, a response with a finite zero is the most useful. I modified the values in Figure 3.9(a) to $R = 39$ kΩ, $C = 5n6$, $C_7 = 1n$, $R_{2,3,6,7} = 100$ kΩ, $R_{4,5} = 47$ kΩ and got the response shown in Figure 3.9(b).

If the output of the basic filter (Figure 3.6) is fed back to its input via an inverting amplifier, there will be zero phase shift at 85% of the cut-off frequency, so an oscillator should result. I tried this using a

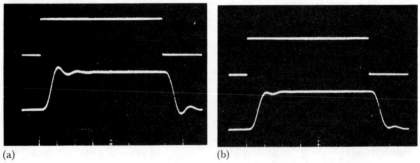

Figure 3.8 Using a lowpass filter to give an approximation to (a) a Butterworth and (b) a Bessel step response

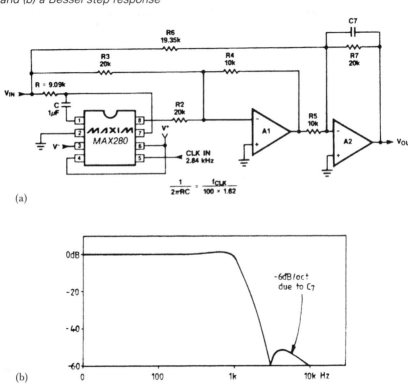

Figure 3.9 Modifying the lospass filter circuit, (a) with R = 39 kΩ, C = 5n6, C_7 = 1 n, $R_{2,3\,6,7}$ = 100 kΩ and $R_{4,5}$ = 47 kΩ gives the response shown in (b)

pair of diodes for amplitude control (Figure 3.10), and got a very convincing looking sinewave. The total harmonic distortion meter indicated 2%, which sounds disappointing, but looking at the residual with the 'scope showed it to consist almost entirely of *SC* switching

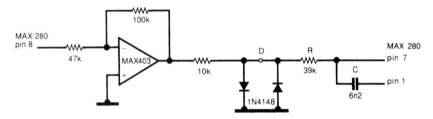

Figure 3.10 *Turning a filter into an oscillator. Feeding the output from the basic filter to its input via an inverting amplifier, using a pair of diodes for amplitude control, gives a good sinewave. Switching the THD meter's bandwidth gives a 'virtually pure' third harmonic*

hash. Switching the THD meter's bandwidth from 80 kHz to 20 kHz gave a more respectable figure of 0.18% THD, virtually pure third harmonic.

Practical considerations

In applying the *MAX280*, a number of practical points arise. If only the ac component of the signal is of interest, the output can be taken from the buffered low impedance output at pin 8, but if the dc component is also important note that there may be an offset of up to 2 mV. In this case, use the dc accurate 'output' at pin 7, which is connected directly via R to the filter's input – the buffer's typical input bias current of 2 pA is not likely to drop a significant voltage across R. The dc accurate output should still be buffered before feeding to, for example, an A-to-D converter, since the pin 7 to pin 1 path is part of the filter, and capacitive loading of even as little as 30 pF at pin 7 may affect the filter response. A passive RC post filter is also recommended to suppress the 10 mV pp (typical) clock feed-through hash. At the other end of the spectrum, the filter contributes no low frequency or $1/f$ noise, since any such noise in the active circuitry would have to pass from pin 1 to the output via a passive CR highpass filter. For critical filtering applications, two *MAX280*s may be cascaded to provide a tenth-order dc accurate filter.

Amplifiers with ultra high input impedance

High input impedances are required for bridge detector circuits used in measuring small capacitances. An imput impedance of 10 GΩ at dc and up to higher audio frequencies is easily arranged with modern devices.

Bootstrap base to bridge building

Bootstrapping is a powerful technique which has long featured in the circuit designer's armoury. Its invention is often ascribed to A.D. Blumlein, in connection with his pre-war work at the laboratories of EMI developing the 405 line television system. It enabled the signal lead from the TV camera tube to its preamplifier to be screened, without adding so much stray capacitance as to reduce the signal's bandwidth (Figure 3.11). Another application of bootstrapping is in a bridge detector. In Figure 3.12, both inputs of the detector amplifier should have such a high input impedance that even on extreme bridge ratios, for example when measuring very small capacitances,

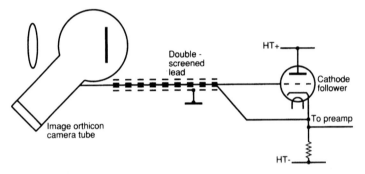

Figure 3.11 *An early application of bootstrapping. The camera signal is connected to its preamplifier via double-screened cable, the inner screen of which is driven by the output of the cathode follower buffer stage. Since the gain of the latter is very nearly unity, there is no ac voltage difference between the inner conductor and the inner of the two screens, so the signal does not 'see' any cable capacitance to ground*

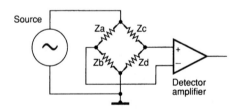

Figure 3.12 *Bridge null detector is a testing application for a differential input amplifier. Both inputs must be very high impedance; additionally, the amplifier requires a high CMRR of around 60 dB for a 1% bridge accuracy*

they do not load the bridge arms at all. In the bridge application, additionally, the detector amplifier should also have a very high CMRR (common mode rejection ratio) – this is particularly importance when the impedance of the lower arms of the bridge are much higher than those of the upper arms, since in this case the difference signal that has to be detected rides on a much larger common mode component.

The necessary high input impedance for such applications is readily achieved using bootstrapping, given a suitable circuit design. A good way to see how effective it is, is to view a squarewave source via a high series resistance. This provides a quick guide as to whether the bootstrapping is effective over a range of frequencies. A very high input impedance at 0 Hz, i.e. a high input resistance, is provided by any JFET input or CMOS opamp; for instance the RCA BiMOS *CA3130* opamp which has been around since the 1970s features a typical input resistance of 1.5 TΩ or 1.5×10^{12} Ω. The input characteristics of some of the wide range of Texas Instruments opamps is shown in Table 3.1.

Table 3.1 *Input characteristics of TI opamps*

Device	C_{in} (pF)	I_{bias} (typ. at 25 kΩ)	R_{in} typ.
TLC27L9	not quoted	0.6 pA	1 TΩ
TLC2201	not quoted	1 pA	not quoted
TLE2021 (bipolar)	not quoted	25 nA	not quoted
TLE2027, TLE2037	8 pF	15 nA	not quoted
TLE2061, TLE2161	4 pF	3 pA	1 TΩ

TLE2061 attraction

The *TLE2061* is a good choice to experiment with, because of its low input capacitance and high input impedance. This JFET input micropower precision opamp offers a high output drive capability of ±2.5 V (min.) into 100 Ω on ±5 V rails and ±12.5 V (min.) into 600 Ω on ±15 V rails, while drawing a quiescent current of only 290 µA. The device operates from V_{cc} supplies of ±3.5 V to ±20 V, with an input offset voltage as low as 500 µV ($\overline{BC}$ version), whilst its decompensated cousin, the *TLE2161*, features an enhanced slew rate of 10 V/µs for applications where the closed loop gain is ×5 or more. The *TLE2061* was connected as a unity gain non-inverting buffer (Figure 3.13), and a 1 kHz 0 V to +4 V squarewave input (upper trace) applied. The spikes on the leading edges appear to be an artefact of the digital storage oscilloscope's screen dump software, there being no trace of them on the oscilloscope trace. Allowing for that, the opamp's output (lower trace) is pretty well a perfect replica, as would be expected. Next, a 10 MΩ resistor was placed in series with the opamp's input (Figure 3.14), the input and output then appearing as in the upper and lower traces respectively. With the

112 Analog circuits cookbook

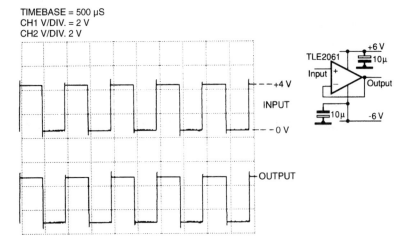

Figure 3.13 *JFET input unity gain buffer circuit's output is indistinguishable from its input*

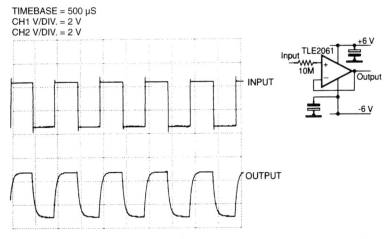

Figure 3.14 *A series 10 MΩ resistor has no effect on the peak-to-peak amplitude, but grossly limits the high frequency response*

opamp's 4 pF input capacitance and allowing 1 pF for strays, the input circuit time constant comes to 50 μs, and viewing the lower trace at a faster timebase speed showed that the time to 63% response was indeed just 50 μs. Clearly in this application, the influence of the input capacitance is far more significant than that of the input resistance. Use of guard rings as recommended in the data sheet will maintain the high input resistance and will minimise stray capacitance external to the opamp (Figure 3.15); the similarity to Figure 3.11 is clear.

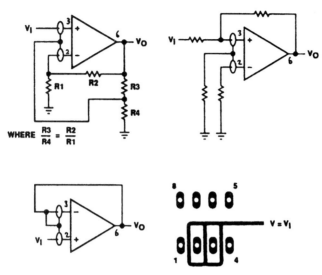

Figure 3.15 *Guard rings around the input terminals minimise the effect of board leakage and capacitance by surrounding the input pins with copper track at the same potential as the input. There is thus no potential difference to force current through any leakage paths or through stray capacitance*

Bootstrapping, however, cannot reduce the effect of the device's internal input capacitance. So my next experiment preceded the opamp with a discrete bipolar buffer stage, using a *BC108* (Figure 3.16). The inadequate input resistance and lower than unity gain with this arrangement is evident on comparing the lower trace with the upper, but the high frequency response is better than in Figure 3.14 – which is to be expected as the input capacitance is now only that of the transistor, mainly C_{cb} or C_{obo} approximately. The data sheets give this as 6 pF max., although my ancient *Transistor DATA Book* (Vol. 1, 1977) gives C_{cb} typical as 2.5 pF. The input time constant is about half that in the circuit of Figure 3.14.

Bootstrapping boon

On the face of it, the result is hardly an improvement; slightly lower input capacitance has simply been traded for a much lower input

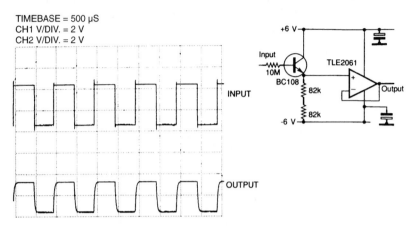

Figure 3.16 *A discrete emitter follower buffer ahead of the opamp improves the high frequency response by a factor of two, but low input resistance pulls down the peak-to-peak output*

resistance. But this is where bootstrapping really comes into its own, hauling the input up by its own bootstraps.

Stage 1 involves bootstrapping the *BC108*'s collector (Figure 3.17), which is seen to be very effective indeed in shortening the input time constant, though there is still a shortfall in low frequency gain. The all-important point to note is that the bootstrapping of the collector only works because there is a separate stage following the emitter follower, providing current gain. The collector cannot be bootstrapped from the input emitter follower's own emitter even though such an arrangement

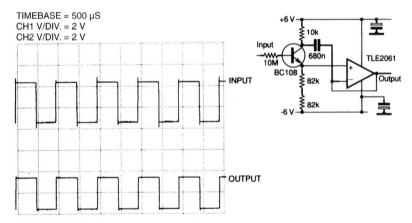

Figure 3.17 *Bootstrapping the emitter follower's collector shortens the input time constant to a negligible value, but the dc gain is still well below unity*

was seriously proposed in *Wireless World* by a well-known writer on electronics, whose name shall remain unstated to spare his blushes.

Stage 2 extends the bootstrapping to the input emitter follower's emitter circuit (Figure 3.18), and now the output (lower trace) is indistinguishable from the input. However, the improvement does not extend down to dc, the input resistance at 0 Hz being unchanged, but only down to a frequency where the time constant of the emitter bootstrapping circuit starts to be significant.

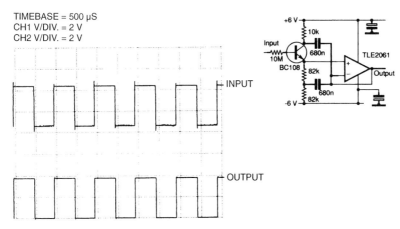

Figure 3.18 *Bootstrapping the emitter circuit as well results in an indistinguishable output*

To extend the bootstrapping down to dc, the emitter circuit bootstrap capacitor would need to be replaced by a zener diode. To see how far it was possible to push the circuit, I replaced the 10 MΩ input resistor by a string of five 10 MΩ resistors in series. The result was the substantially reduced output shown in Figure 3.19 – an unduly rapid collapse in performance, bearing in mind how good the performance was with 10 MΩ series resistance. Probing around the circuit showed that the emitter swing was between –2 V and –4 V, due to the volt drop caused by the transistor's base current flowing through the 50 MΩ resistor, with the result that the emitter current was totally inadequate. This reminded me of the old adage: when your circuit isn't behaving as you think it ought, check the dc conditions.

Raising the opamp supply rails to ±15 V resulted in an output virtually as good as in Figure 3.18. Clearly, properly applied, bootstrapping can raise the input impedance at dc and up to a frequency determined by the opamp's performance, to such a high level that a 100 MΩ source resistance results in no loss in amplitude, i.e. to an input impedance of 10 GΩ or more.

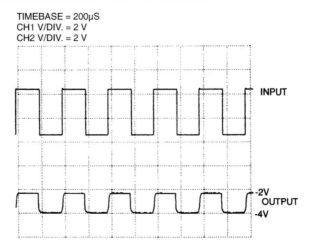

Figure 3.19 *As Figure 3.18, but with the input resistor raised to 50 MΩ. The poor performance is the fault of the designer, not the circuit*

As a matter of interest, the circuit of Figure 3.18 (with ±15 V rails) can be modified by substituting a *BF244* N-channel small signal JFET for the *BC108*. Now, of course, there is no volt drop across the 50 MΩ input resistance, and the opamp output voltage sits at a positive level set by the FET's gate source reverse bias voltage at the source current defined by the two 82 kΩ resistors. The FET's drain gate capacitance is the best part of 2 pF, so the collector bootstrapping is still necessary. However, the input resistance is so high that a source circuit bootstrapping capacitor is not needed.

> ### Some integrated active filters
> Using integrated filter packages has never been easier. This article describes their application, and an audio circuit to test the response.

Mighty filter power in minuscule packages

Although digital electronics still hogs much of the limelight, analog electronics continues to advance, quietly but steadily. Indeed, if the renewal of interest in rf, due to all the various developments afoot in the personal communications scene, is included – rightly – under the generic heading 'analog', then some semblance of balance between the two halves of the great divide has re-established itself. The ICs which are introduced below are typical of the increasing power and sophistication available in analog electronic devices. Having obtained some samples, I set about exploring their capabilities.

The Maxim devices *MAX291–MAX297* are eighth-order lowpass switched capacitor filters available in 8 pin plastic DIP, SO, CERDIP packages, 16 pin wide SO packages, and even chip form. They cover a variety of filter types, namely Butterworth, Bessel, elliptic (minimum stopband attenuation $A_s = 80$ dB from a stopband frequency F_s of 1.5 × the corner (cut-off) frequency F_o) and elliptic (A_s 60 dB at $1.2 \times F_o$). The corresponding type numbers are *MAX291/292/293/294* respectively, all at a ratio of clock to corner frequency of 100:1. The *295/296/297* are Butterworth, Bessel and elliptic (A_s 80 dB) types, but employing a 50:1 clock ratio, extending the maximum F_o to 50 kHz, against 25 kHz for the others. All will accept an external clock frequency input, enabling the corner frequency to be accurately determined and to be changed at will, or can be run using an internal clock oscillator, the frequency of which is determined by a single external capacitor. Whilst typical frequency response curves are given in the data sheets, it would clearly be an interesting exercise to measure the responses independently, for which purpose an audio swept frequency source and detector are called for. The simple arrangement of Figure 3.20(a) was therefore constructed.

Figure 3.20(b) shows the result of applying the swept output direct to the detector. The low amplitude at low frequencies is in fact due to two separate effects. Firstly, at low frequencies the output impedance of the internal current sources and the input impedance of the internal simple Darlington buffers in IC_2 are not infinitely large compared with the reactance of the 1.5 nF capacitors. The second effect is the rate of change of frequency, which at the start of the ramp is comparable to the actual output frequency itself. The purpose of this was to allow the individual cycles of the frequency ramp to be seen. For measuring the filter responses, a much slower ramp would clearly be necessary – to enable the detector to follow rapid downward changes in level – so this second effect would not apply. However, the first still would, but for the current purpose – it was intended to operate the filters at a 1 kHz cut-off frequency – this was of no consequence.

For testing the frequency responses of the filters, the value of C was raised from 1 nF to 680 nF, giving a sweep time of one minute. At this slow rate, the limited dot density of the digital storage oscilloscope resulted in a ragged meaningless depiction of the swept frequency test signal itself. Figure 3.21(a) therefore shows the sweep voltage instead (upper trace), together with the detected output from the filter (lower trace, taken using the *MAX291* Butterworth filter).

118 *Analog circuits cookbook*

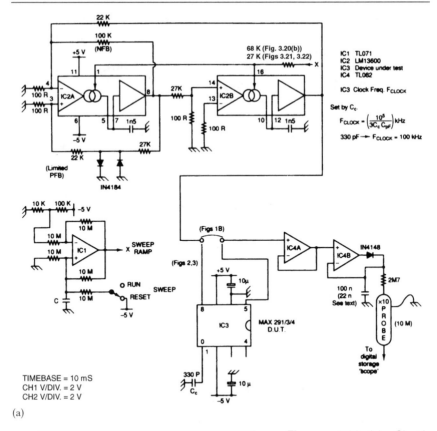

(a)

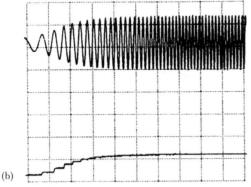

(b)

Figure 3.20 (a) Simple audio swept frequency response measurement system. A Howland current pump is used to charge capacitor C, providing a linear sweep voltage at the output of opamp IC_1. This is applied to the bias inputs of a LM13600 dual operational transconductance amplifier (OTA, IC_2), used as a voltage-controlled state-variable-filter based sinewave oscillator. Its output is applied to the device under test, IC_3, the output of the latter being detected by the ideal rectifier circuit IC_4. (b) Using a small value of C, the swept oscillator output was applied direct to the detector circuit. The detected output (lower trace) follows faithfully the peak amplitude of the sweeper output (upper trace) over the partial scan shown, covering about 30 Hz to 650 Hz

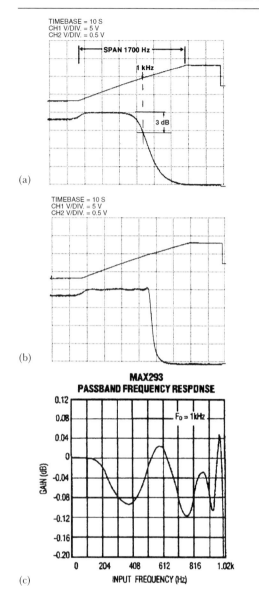

Figure 3.21 (a) The ramp-voltage applied to the swept frequency oscillator (upper trace) and the detected voltage output from the MAX291 Butterworth 8 pole filter, set to F_o = 1 kHz (lower trace). (b) As (a), but using the MAX293 elliptic filter with its 1.5:1 ratio of F_s to F_o. (c) The manufacturer's frequency response data for the MAX293

The amplitude of the sinewave test signal settles rapidly to about 5 V pp at the start of the sweep and remains constant over the whole sweep, whilst the detected output starts to fall at the filter's corner frequency, being as expected 3 dB down at 1 kHz. (The detected voltage is 2 V, not 2.5 V, due to the attenuation introduced at the trace 2 probe, to avoid overloading the digital storage oscilloscope's channel 2 A-to-D converter; the alternative of reducing the sensitivity from 0.5 V/div. to 1 V/div. would have resulted in rather a small deflection.) The 3 dB attenuation at F_o and leisurely descent into the stopband, typical of the maximally flat Butterworth design, are clearly shown. Contrast this with the MAX293 A_s = 80 dB elliptic filter (Figure 3.21(b)), which has dropped by 20 dB from the passband level within a space of around 200 Hz. The maker's data (Figure 3.21(c)) shows the gain variations in the passband, on a much expanded scale.

Using the linear detector shown in Figure 3.20, it is not of course possible to see in Figure 3.21(b) the detail of the stopband. Detail up to

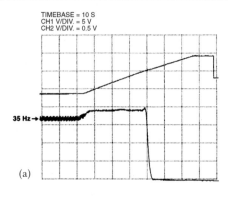

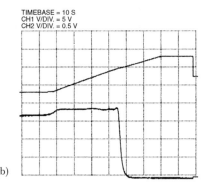

Figure 3.22 *As Figure 3.21(a), but using the MAX294 elliptic filter with its 1.2:1 F_s to F_o ratio, using a modified detector circuit. (b) As (a), but the detector circuit as in Figure 3.20*

around 80 dB down would be visible using the logamp circuit described in Chapter I, 'Logamps for radar – and much more' (see also Hickman, 1993) but this would still be insufficient to examine the stopband of this particular device adequately. It would, however, be adequate for viewing the stopband detail of the MAX294, the performance of which in the set-up of Figure 3.20 is shown in Figure 3.22(a): the minimum stopband attenuation of 60 dB offered by this device is maintained whilst providing an F_s to F_o ratio of only 1.2:1. This plot was taken with the smoothing capacitor in IC_4's linear detector circuit reduced from 100 nF to 22 nF, enabling the detector to follow the very rapid cut-off of the filter at the given sweep speed. Accordingly, increased ripple is observable on the detector output as low frequencies preceding the start of the sweep.

Figure 3.22(b) shows the same response with the original detector time constant, showing the distorted response caused by using an excessive post-detection filter time constant – a point which will not be lost upon anyone who has used early spectrum analysers which did not incorporate interlocking of the sweep speed, span, IF bandwidth and post-detector filter settings. The measurement could of course have been taken without error using the original detector by reversing the polarity of the ramp to give a falling frequency test signal – at the expense of having a back-to-front frequency base. Conversely, there would be no problem with the original arrangement when measuring a highpass filter, since the detector's response to increasing signals is very fast. The design of a detector with low output ripple but with fast reponse to both increasing and decreasing signal levels is an interesting exercise.

The maximally flat Butterworth response of Figure 3.21(a) is of course free of peaking, but peaking can be expected in the elliptic responses. In Figure 3.21(b) it appears to be about 1% at F_o, corresponding to +0.086 dB. This is within the maker's tolerance, also measured at 1 kHz, which is –0.17 to +0.12 dB, with +0.05 dB being typical. With the faster cut-off offered by the *MAX294*, somewhat larger peaking (–0.17 to +0.26) is to be expected, and is observed (Figure 3.22(a)). Note that measurement accuracy is limited by many factors other than the detector time constant mentioned above. For instance, the distortion of the sinewave test signal produced by IC_2, measured at 1 kHz, is as much as 0.6%. It consists almost entirely of third harmonic, which is thus only 44 dB down on the fundamental. Even assuming that the level of the latter is exactly constant over the sweep, using a peak detector circuit a 0.05 dB change in level can be expected at 333 Hz, at which point the third harmonic sails out of the filter's passband. Thus a very clean, constant amplitude test signal indeed would be necessary to test the filter's passband ripple accurately. It would also be necessary for even basic measurements on a highpass filter, where the harmonic(s) of the test signal would sweep into the filter's passband whilst the fundamental was still way down in the stopband.

All the filters in the range offer very low total harmonic distortion (THD), around –70 dB. Consequently the elliptic filters lend themselves very nicely to the construction of a digitally controlled audio oscillator. Such a circuit was constructed and is shown in Figure 3.23(a). The *'LS90* was pressed into service because it will divide by ten whilst giving a 50/50 mark/space ratio output, and also because I had plenty in stock. The $F_{clock}/100$ output of the second *'LS90*, suitably level shifted, was applied to the *MAX294*'s signal input, pin 8, and the clock input itself to pin 1. The *MAX294* will operate on a single +5 V rail (in which case the signal input should be biased at +2.5 V) or, as here, on +5 V and –5 V rails. Either way it will accept a standard 0 to +5 V CMOS clock input at up to 2.5 MHz or, as it turns out in practice, a *74LSXX* input, though this is not stated in the data sheet. The *'LS90* may be old hat, but it is nonetheless fast, so a clean clock drive and local decoupling were used to ensure no false counting due to glitches, etc.

The attenuation of the *MAX294* at $3F_o$ is around 60 dB and bearing in mind that the third harmonic component of the squarewave input to the device is 9.5 dB down on the fundamental, the squarewave should be filtered into a passable sinewave with all harmonics 70 dB or more down. This is comparable in level to the device's stated THD, so that

122 Analog circuits cookbook

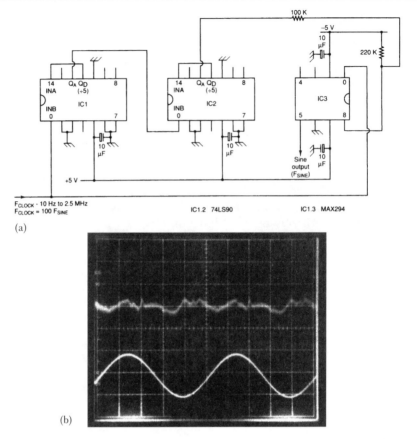

Figure 3.23 *(a) Circuit of a digitally tuned sinewave audio oscillator using the MAX294. (b) The circuit's output at 1 kHz (lower trace) and the residual signal after filtering out the fundamental, representing the total harmonic distortion (upper trace)*

although the *MAX293* could equally well be used in this application, its greater stopband attenuation would not in fact be exploited. The Butterworth *MAX291* also shows greater than 60 dB attenuation at $3F_o$ relative to F_o: at $2F_o$ it is only just over 40 dB relative, but of course the squarewave drive has no second harmonic. Consequently, the *MAX291/293/294* are all equally suitable in this application.

Figure 3.23(b) (lower trace) shows a 1 kHz sinewave output from the circuit in Figure 3.23(a); the 100 kHz steps forming the waveform are just visible. At first sight, it looks very like the waveform out of a DDS (direct digital synthesiser), but there are one or two subtle differences. Timewise, the quantisation is always exactly 100 steps per cycle, whereas in a DDS it can be any number of times (clock frequency divided by maximum accumulator count), the latter being typically 2^{32}.

Considering amplitude, the waveform is simply just not quantised; it is an example of a true PAM system, where each step can take exactly the appropriate value for that point in a continuous sinewave. Figure 3.23(b) also shows the residual THD (upper trace), being the monitor output of a THD meter on the 0.1% FSD range. The measured THD was 0.036% or 69 dB down on the fundamental. This agrees exactly with the manufacturer's data (Figure 3.24(a)), which shows that the level of THD + noise relative to the signal is independent of the actual signal level over a quite wide output range. The slight fuzziness of the THD trace is due to some 50 Hz getting into the experimental lash-up, not (as might be supposed) residual clock hash. The latter was suppressed by switching in the THD meter's 20 kHz lowpass filter: without this necessary precaution the residual signal amounted to just over 1%.

The *MAX29X* series of switched capacitor filters each includes an uncommitted opamp which can be used for various purposes. It makes a handy anti-aliasing filter to precede the main switched capacitor section or can alternatively be used as a post-filter to reduce clock breakthrough in the output. Unfortunately, it cannot suppress it entirely, being part of the same very busy ship as the 8 pole switched capacitor filter section. Its use is illustrated in Figure 3.24(e).

Where a modest distortion figure of somewhere under 0.05% is adequate, an instrument based on the circuit of Figure 3.23(a) has certain attractive features. It can cover 0.1 Hz to 25 kHz with a constant amplitude output and much the same THD over the whole range, given suitable post-filtering to suppress clock hash. The post-filters need to be selected as appropriate, but with a clock frequency of 100 times the output frequency each can cover a 20:1 frequency range or more. This means that only two or three are needed to cover the full 20 Hz to 20 kHz audio range, while four can cover the range 0.1 Hz–25 kHz. The clock can be fed to a counter with a 100 ms gate time, providing near instantaneous digital readout of the output frequency down to 20 Hz to a resolution of 0.1 Hz, a feature which would require a 10 s gate time in a conventional audio oscillator with digital read-out. If the clock is derived from a DDS chip, then the frequency can be set digitally, to crystal accuracy. The clock division ratio of 100 would reduce any phrase-modulation spurs in the output of the DDS by 40 dB: a necessary feature with many DDS devices.

The usual arrangement in a multipole active filter is to cascade a number of individual sections, each of which is solely responsible for one pole pair of the overall response. This can lead to substantial departures from the desired response, due to component tolerances in the individual 2 pole sections, particularly the highest Q section(s). Interestingly, the *MAX29X* series filters employ a design which

124 *Analog circuits cookbook*

(a)
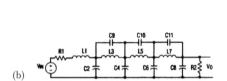

Pin Description

8-PIN	16-PIN	NAME	FUNCTION
	1, 2, 7, 8, 9, 10, 15, 16	N.C.	No Connect
1	3	CLK	Clock input. Use internal or external clock.
2	4	V-	Negative Supply pin. Dual supplies: -2.375V to -5.500V. Single supplies: V- = 0V.
3	5	OP OUT	Uncommitted Op-Amp Output
4	6	OP IN-	Inverting Input to the uncommitted op amp. The noninverting op amp is internally tied to ground.
5	11	OUT	Filter Output
6	12	GND	Ground. In single-supply operation, GND must be biased to the mid-supply voltage level.
7	13	V+	Positive Supply pin. Dual supplies: +2.375V to +5.500V. Single supplies: +4.75V to +11.0V.
8	14	IN	Filter Input

LABEL	f_CLK (Hz)	F_o (kHz)	INPUT FREQ. (Hz)	MEASM^T BANDWIDTH
A	200k	2	200	30 KHz
B	1M	10	1k	80 KHz

(b)
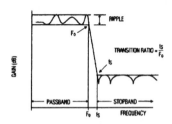

(c)
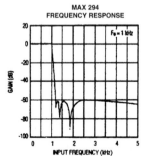
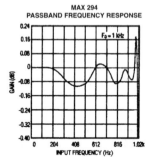
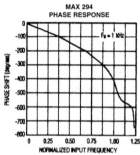

Measurements (audio and video) 125

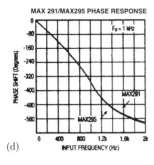
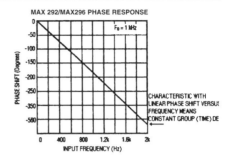

(d)

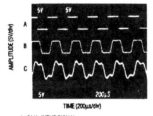

A: 3 kHz INPUT SIGNAL
B: MAX292 BESSEL FILTER RESPONSE WITH Fo=10 kHz
C: MAX 291 BUTTERWORTH FILTER RESPONSE WITH Fo=10 kHZ

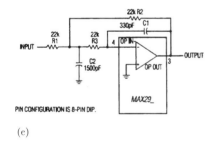

(e)

PIN CONFIGURATION IS 8-PIN DIP.

Table 1. Component values.

Corner frequency (Hz)	R_1 (kΩ)	R_2 (kΩ)	R_3 (kΩ)	C_1 (F)	C_2 (F)
100k	10	10	10	68p	330p
50k	20	20	20	68p	330p
25k	20	20	20	150p	680p
10k	22	22	22	330p	1.5n
1k	22	22	22	3.3p	15n
100	22	22	22	33n	150n
10	22	22	22	330n	1.5μ

Note: some approximations have been made in selecting preferred component values.

The passband error caused by a 2nd-order Butterworth can be calculated using the formula

$$\text{Gain error} = -10\log\left[1 + \left(\frac{f}{f_c}\right)^4\right] \text{dB}$$

Figure 3.24 (a) THD + noise relative to the input signal amplitude for the MAX294. (b) The MAX29X series filter structure emulates a passive 8 pole lowpass filter. In the case of the elliptic types, this results in ripples in both the pass- and stopbands. (c) Passband and stopband performance for the MAX294 with a 100 kHz clock (F_o = 1 kHz). (d) Comparison of the pulse response of the Bessel and Butterworth filter types. (e) Use of the MAX29X's uncommitted opamp as an aliasing filter

emulates a passive ladder filter (Figure 3.24(b)), so that any individual component tolerance error marginally affects the shape of the whole filter rather than being concentrated on a particular peak. Ideally, the passband peaks and troughs are all equal, as are the stopband peaks. The actual typical performance (for the *MAX294*) is shown in Figure 3.24(c).

The Butterworth filter (with simple pre- and post-filters) provides a powerful and anti-aliasing function to precede the A-to-D converter of a DSP (digital system processor) system. The elliptic versions enable operation even closer to the Nyquist rate (half A-to-D's sampling frequency), the *MAX294* being suitable for 10-bit A-to-Ds and the *MAX293* for 12 or 14 bit A-to-Ds. This assumes that the following DSP system is interested only in the relative amplitudes of the frequency components of the input, and not in their relative phases. Where the latter is also important, to preserve the detailed shape of the input, the *MAX292* filter with its Bessel response is needed. Alias-free operation will then be possible only to a lower frequency; e.g. one-fifth of the Nyquist rate for a 10 bit system, since $A_s = 60$ dB occurs at $5F_o$ for this device. However, compared with a Butterworth filter, the improved waveform fidelity of the Bessel filter with its constant group delay is graphically illustrated in Figure 3.24(d). The pulse response of the elliptic types would be even more horrendous than the Butterworth.

To perform in DSP the same filtering function as provided by a *MAX29X* would require much greater expenditure of board space, power, money and number of chips. These devices provide mighty filter power in minuscule packages.

References

Hickman, I. (1993) Logamps for radar – and much more. *Electronics World + Wireless World*, April, 314–317.

'Scope probes – active and passive

Extending oscilloscope measurement capability.

Introduction

An oscilloscope is the development engineer's most useful tool – it shows him what is actually going on in a circuit. Or it should do, assuming that connecting the oscilloscope to a circuit node does not change the waveform at that node. To ensure that it doesn't, oscilloscopes are designed with a high input impedance. The standard value is 1 MΩ, in parallel with which is inevitably some capacitance, usually about 20–30 pF.

As far as the power engineer working at mains frequency is concerned, this is such a high value as to be safely ignored, and the same goes for the audio engineer – except, for example, when examining the early stages of an amplifier, where quite high impedance nodes may be encountered. But the 'scope's high input impedance exists at its input socket, to which the circuit of interest must be connected. So some sort of lead is needed – connecting a circuit to an oscilloscope with leads of near zero length is always difficult and tedious, and often impossible. Sizeable low frequency signals emanating from a low impedance source present no difficulty, any old bit of bell flex will do. But in most other cases a screened lead will be needed, to avoid pick-up of hum or other extraneous signals.

A screened lead of about a metre or a metre and a half proves to be convenient, and such a lead would add somewhere between 60 and 150 pF of capacitance to that at the 'scope's input socket. But the reactance of just 100 pF at even a modest frequency such as 1 MHz is as low as 1600 Ω, a far cry from 1 MΩ and not generally negligible by any stretch of the imagination. The usual solution to this problem is the 10:1 passive divider probe. This provides at its tip a resistance of 10 MΩ in parallel with a capacitance of around 10 pF; not ideal, but a big improvement over a screened lead, at least as far as input impedance is concerned. But the price paid for this improvement is a heavy one, the sensitivity of the oscilloscope is effectively reduced by a factor of ten.

Passive divider probes

Figure 3.25(a) shows the circuit of the traditional 10:1 divider 'scope probe, where C_O represents the oscilloscope's input capacitance, its input resistance being the standard value of 1 MΩ. The capacitance of the screened lead C_C plus the input capacitance of the 'scope form one section of a capacitive potential divider. The trimmer C_T forms the other, and it can be set so that the attenuation of this capacitive divider is 10:1 in volts, which is the same attenuation as provided by R_A (9 MΩ) and the 1 MΩ input resistance of the oscilloscope. When this condition is fulfilled, the attenuation is independent of frequency – Figure 3.26(a). Defining the cable plus 'scope input capacitance as C_E, i.e. $C_E = C_C + C_O$ (Figure 3.25(b)), then C_T should have a reactance of nine times that of C_E, i.e. $C_T = C_E/9$. If C_T is too small, high frequency components (e.g. the edges of a squarewave) will be

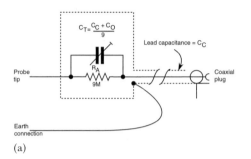

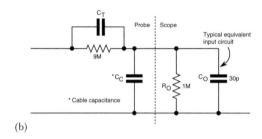

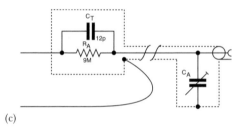

Figure 3.25 (a) Circuit of traditional 10:1 divider probe. (b) Equivalent circuit of probe connected to oscilloscope. (c) Modified probe circuit with trimmer capacitor at 'scope end

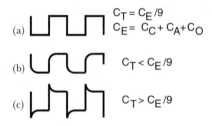

Figure 3.26 Displayed waveforms with probe set up (a) correctly, (b) undercompensated, (c) overcompensated

attenuated by more than 10:1, resulting in the waveform of Figure 3.26(b). Conversely, if C_T is too large, the result is as in Figure 3.26(c).

The input capacitance of an oscilloscope is invariably arranged to be constant for all settings of the Y input attenuator. This means that C_T can be adjusted by applying a squarewave to the 'scope via the probe using any convenient Y sensitivity, and the setting will then hold for any other sensitivity.

The circuit of Figure 3.25(a) provides the lowest capacitive circuit loading for a 10:1 divider probe, but has the disadvantage that 90% of the input voltage (which could be very large) appears across the variable capacitor C_T. Some probes therefore use the circuit of Figure 3.25(c): C_T is now a fixed capacitor and a variable shunt capacitor C_A is fitted, which can be set to a higher or lower capacitance to compensate for 'scopes with a lower or higher input capacitance respectively. Now, only 10% of the input voltage appears across the trimmer, which is also conveniently located at the 'scope end of the probe lead, permitting a smaller, neater design of probe head.

Even if a 10:1 passive divider probe (often called, perhaps confusingly, a ×10 probe) is incorrectly set up, the rounding or pip on the edges of a very low frequency squarewave, e.g. 50 Hz, will not be very obvious, because with the slow timebase speed necessary to display several cycles of the waveform, it will appear to settle instantly to the positive and negative levels. Conversely, with a high frequency squarewave, say 10 MHz, the probe's division ratio will be determined solely by the ratio C_E/C_T. Many a technician, and chartered engineer too, has spent time wondering why the amplitude of a clock waveform was out of specification, only to find eventually that the probe has not been set up for use with that particular oscilloscope. Waveforms as in Figure 3.26 will be seen with a squarewave of around 1 kHz.

Probe behaviour at high frequencies

At very high frequencies, where the length of the probe lead is an appreciable fraction of a wavelength, reflections would occur, since the cable is not terminated in its characteristic impedance. For this reason, oscilloscope probes often incorporate a resistor of a few tens of ohms in series with the inner conductor of the cable at one or both ends, or use a special cable with an inner made of resistance wire. Such measures are necessary in probes that are used with oscilloscopes having a bandwidth of 100 MHz or more.

Whilst a 10:1 passive divider probe greatly reduces the loading on a circuit under test compared with a similar length of screened cable, its effect at high frequencies is by no means negligible. Figure 3.27 shows the typical variation of input impedance versus frequency of such a probe, when connected to an oscilloscope. Another potential problem area to watch out for when using a 10:1 divider probe is the effect of the inductance of its ground lead. This is typically 150 nH (for a 15 cm lead terminated in a miniature 'alligator' clip), and can form a resonant circuit with the input capacitance of the probe. On fast edges, this will result in ringing in the region of 150 MHz, so for high frequency applications it is essential to discard the ground lead and to earth the grounded nose-ring of the probe to circuit earth by the shortest possible route.

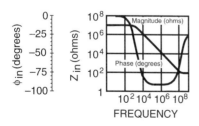

Figure 3.27 *Variation of impedance with frequency at the tip of a typical 10:1 passive divider probe (Courtesy Tektronix UK Ltd)*

Active probes

Figure 3.27 shows that over a broad frequency range – say roughly 30 kHz to 30 MHz – the input impedance of a 10:1 passive divider probe is almost purely capacitive, as evidenced by the almost 90° phase angle. But it can be seen that at frequencies well beyond 100 MHz, the input impedance of the probe tends to 90 Ω resistive – the characteristic impedance of the special low capacitance cable used. At frequencies where C_T is virtually a short circuit, the input of the probe cable is connected directly to the circuit under test, causing heavy circuit loading.

The only way round this is to fit a buffer amplifier actually in the probe head, so that the low output impedance of the buffer drives the cable, isolating it entirely from the circuit under test. Such active probes have been available for many years for top-of-the-line oscilloscopes from the major manufacturers, and in many cases, their oscilloscopes are fitted with appropriate probe power outlets. Figure 3.28 shows the circuit diagram of such an active probe, the Tektronix *P6202A* providing a 500 MHz bandwidth and an input capacitance of 2 pF, together with stackable clip-on caps to provide ac coupling or an attenuation factor of ten to increase the dynamic range. The circuit illustrates well how, until comparatively recently, when faced with the need to wring the highest performance from a circuit, designers were still forced to make extensive use of discrete components. Note that such an active probe provides two important advantages over the passive 10:1 divider probe. Firstly, the input impedance remains high over the whole working frequency range, since the circuit under test is buffered from the low impedance of the output signal cable. Secondly, the factor of ten attenuation of the passive probe has been eliminated.

Whilst high performance active probes are readily available, at least for the more expensive models of oscilloscope, their price is high. The result is that most engineers are forced to make do, reluctantly, with passive probes, with their heavy loading (at high frequencies) on the circuit under test, and the attendant loss of a factor of ten in sensitivity. Whilst passive divider probes (at affordable prices) for oscilloscopes with a bandwidth of 60 to 100 MHz are readily available, active probes of a similar modest bandwidth are not. But with the continuing improvements in opamps of all sorts, it is now possible to design simple active probes without resorting to the complexity of a design using discretes such as Ref. 1 or Figure 3.28.

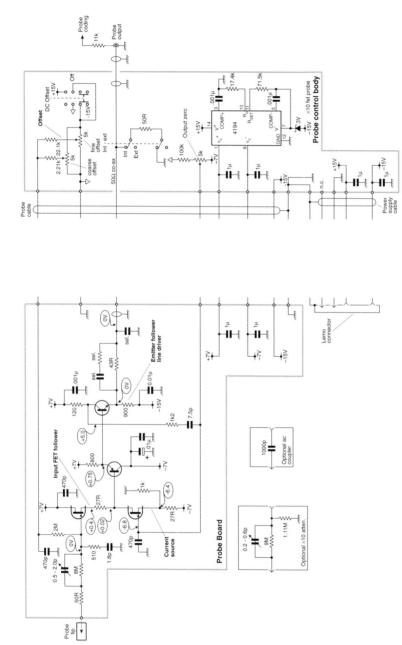

Figure 3.28 Circuit diagram of the P6202A active FET input probe, with a dc – 500 MHz bandwidth and 2 pF input capacitance (Courtesy Tektronix UK Ltd)

Some active probes

To provide a 10 MΩ input resistance, the same as a passive 10:1 divider probe, an active probe built around an opamp must use a MOS input type. For optimum performance at high frequencies, it is desirable that the opamp should drive the coaxial cable connecting the probe to the oscilloscope as a matched source, so that in the jargon of the day, the cable is 'back-terminated'. This, together with a matched termination at the 'scope end of the probe lead, will divide the voltage swing at the output of the opamp by two. So for a unity gain probe, the opamp must provide a gain of ×2. For this purpose, an opamp which is partially decompensated, for use at a gain of two or above, is very convenient. An active probe using such a MOS-input opamp, the SGS-Thomson *TSH131*, is shown in Figure 3.29(a). This opamp has a 280 MHz gain-bandwidth product, achieved by opting for only a modest open loop gain; the large-signal voltage gain A_{vd} (V_o = ±2.5 V, R_l = 100 Ω) being typically ×800 or 58 dB. At a gain of ×2 it should therefore provide a bandwidth approaching 140 MHz. Care should be taken with the layout to minimise any stray capacitance from the non-inverting input, pin 2, to ground, since this would result in HF peaking of the frequency response. If need be, a soupçon of capacitance can be added in parallel with the 1 kΩ feedback resistor from pin 6, to control the settling time.

A zero offset adjustment is shown, but in most cases this will be judged superfluous. Even with a device having the specified maximum input bias current I_{ib} of 300 pA, the offset due to the 10 MΩ ground return resistor at pin 3 is only 3 mV, whilst the typical device I_{ib} is a meagre 2 pA. With the omission of the offset adjust circuitry, the circuit can be constructed in a very compact fashion on a few square centimetres of copper-clad laminate or 0.1″ matrix strip board, with the output signal routed via miniature 50 Ω coax. The supply leads can be taped alongside the coax to a point near the 'scope end of the probe, where they branch off, allowing a generous length for connection to a separate ±5 V supply, assuming such is not available from the oscilloscope itself. Note the use of a commercially available 50 Ω 'through termination' between the oscilloscope end of the probe signal lead and the Y input socket of the oscilloscope itself.

For ac applications, where it is desired to block any dc level on which the signal of interest may be riding, a blocking capacitor can be incorporated in a clip-on cap to fit over the probe tip. A similar arrangement can be made to house a 10:1 divider pad, to extend the dynamic range of the unit. Without such a pad, the maximum signal that can be handled is clearly quite limited. Bear in mind that ±2.5 V peak-to-peak at the output of the opamp will provide the oscilloscope input with only ±1.25 V, so an attenuator cap will be needed if looking at, for example, clock pulses. But for this purpose, a conventional 10:1 passive divider probe will usually suffice: where an active probe scores is when looking at very small signals, which are too small to measure with a 10:1 passive divider probe. Another application where an active probe scores is when looking at high frequency signals emanating from a high impedance source. Clearly the heavy damping imposed by a passive divider probe at 100 MHz and above precludes its use to monitor the signal across a tuned circuit, whereas the active probe will provide much reduced damping, in addition to enabling much smaller signals to be seen.

An active probe to the circuit of Figure 3.29(a) was made up and tested. As miniature 1/16 W 1K resistors were not to hand, 1.2K resistors were used instead. This, together with the use of a DIL

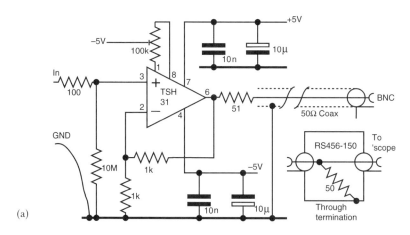

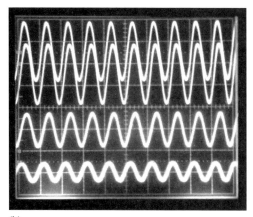

Figure 3.29 (a) Circuit of a unity gain active FET input probe, using a decompensated opamp designed for use at gains of ×2 or greater. Bandwidth should be well over 100 MHz. (b) Performance of the active probe, compared with a P6106 passive probe. 100 mV rms 100 MHz CW output of a signal generator, viewed at 100 mV/div., 10 ns/div. Top and third trace, active probe without and with respectively a 470 Ω resistor in series with tip. Second and bottom trace; same but passive probe

packaged amplifier (in a turned pin socket) rather than the small outline version, meant that some capacitance between pins 2 and 6 was needed. A 0.5–5 pF trimmer was used: it was adjusted so that the probe's response to a 5 MHz squarewave with fast edges was the same as a Tektronix *P6106* passive probe, both being used with a Tektronix *475A* oscilloscope of 250 MHz bandwidth. The advantages of an active probe are illustrated in Figure 3.29(b), where all traces are effectively at 100 mV/div., allowing for the unity gain of the active probe, and the 20 dB loss of the passive probe. All four traces show the 100 MHz CW output of an inexpensive signal generator, the Leader Model *LSG-16*. The measurements were made across a 75 Ω termination, the top trace being via the active probe and the next one via a P6106 passive probe. Both show an output of about 280 mV peak-to-peak, agreeing well with the generator's rated output of 100 mV rms. The third trace shows the same signal, but with a 470 Ω resistor connected in series with the tip of the active probe, whilst the bottom trace is the same again but with the 470 Ω resistor connected in series with the tip of the passive probe.

The effect of the 470 Ω resistor has been to reduce the response of the passive probe by 12 dB, whilst that of the active probe is depressed by only 4.5 dB. Thus the active probe not only provides 20 dB more sensitivity than the passive probe, but exhibits a substantially higher input impedance to boot.

An active probe can be designed not merely to provide unity gain, avoiding the factor of 10 attenuation incurred with a passive divider probe, but actually to provide any desired gain in excess of unity. Figure 3.30(a) shows a circuit providing a gain of ×10, which as before requires a gain of twice that from the opamp. Again, in the interests of providing the conventional 10 MΩ probe input resistance, a FET input opamp was chosen, in this case the Burr-Brown *OPA655*. This device is internally compensated for gains down to unity, and provides a 400 MHz gain-bandwidth product. In this application it is required to provide a gain of ×20, so clearly a decompensated version would provide improved performance. But despite persistent rumours of the imminent appearance of such a version, I have not managed to get my hands on one. At a gain of ×20 or 26 dB, the *OPA655* might be expected to provide a bandwidth of 400/20 or approaching 20 MHz, but note that as more and more gain is demanded of a unity-gain compensated voltage feedback opamp, the bandwidth tends to reduce rather faster than pro rata to the increase in gain.

Figure 3.30(b) records the performance of the ×10 gain active probe of Figure 3.30(a), tested with a 100 mV peak-to-peak 5 MHz squarewave. The rise and fall times of the test squarewave were 4 ns, and of the oscilloscope 1.4 ns. The smaller waveform is the 100 mV

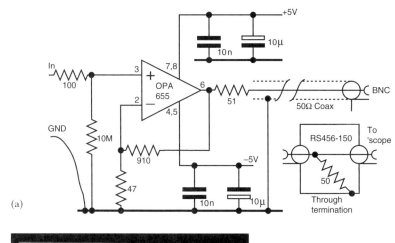

(a)

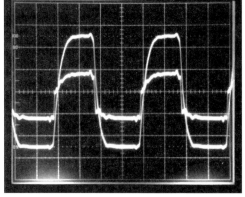

(b)

Figure 3.30 (a) Circuit diagram of an active FET input probe providing a net gain of ×10. (b) 5 MHz 100 mV test squarewave input (smaller trace, at 50 mV/div.), 1 V peak-to-peak output at 'scope (larger trace, at 200 mV/div., at 50 ns/div.

squarewave recorded with a passive 10:1 divider probe with the oscilloscope set to 5 mV/div., effectively 50 mV/div. allowing for the probe. The larger waveform is the 1 V peak to peak output of the active probe, recorded at 200 mV/div. The rise and fall times of the active probe output are 25 ns and 20 ns respectively; it is not uncommon to find differing rise and fall times in high performance opamps, though here the result is influenced also by the shape of the positive-going edge of the test waveform. Taking an average of 22.5 ns and reducing this to 22 ns to allow for the risetimes of the oscilloscope and test waveform, gives an estimated bandwidth for the active probe of 16 MHz, using the formula risetime $t_r = 0.35/BW$, t_r in microseconds, bandwidth BW in MHz. Thus this probe would be useful with any oscilloscope having a 20 MHz bandwidth, the 'scopes' 17.5 ns rise time being increased to 28 ns by the probe.

A much faster probe with a gain of ten can be produced using that remarkable voltage feedback opamp, the Comlinear *CLC425*, which is a decompensated type, for use at gains of not less than ×10. This device is an ultra low noise wideband opamp with an open loop gain of 96 dB and a gain-bandwidth product of 1.7 GHz. At the required gain of ×20 therefore, it should be possible to design an active probe with a bandwidth approaching 85 MHz.

The circuit of Figure 3.31(a) was made up and tested using a 5 MHz squarewave with fast edges, produced with the aid of 74AC series chips, as shown in Figure 3.33(a). The result is shown in Figure 3.31(b), where the smaller waveform is the attenuated test waveform viewed via a 10:1 passive divider probe at 50 mV/div. The test waveform was intended to be 50 mV, but the accumulated pad errors resulted in it actually being 55 mV. The larger trace is the 550 mV

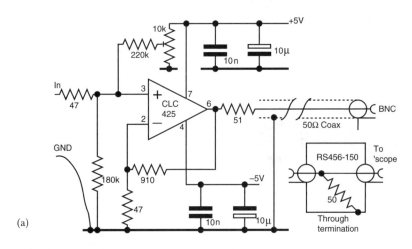

(a)

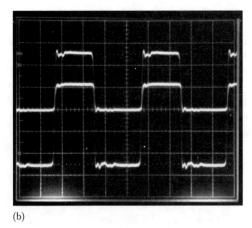

(b)

Figure 3.31 (a) Circuit diagram of an active bipolar probe providing a net gain of ×10. (b) 5 MHz 55 mV test squarewave input (smaller trace, at 50 mV/div.), 550 mV peak-to-peak output at 'scope (larger trace, at >100 mV/div.), at 50 ns/div. Output rise- and falltimes (measured at 10 ns/div., not shown) are 4.5 and 4.0 ns respectively

output from the ×10 active probe, recorded at 100 mV/div. with the oscilloscope's VARiable Y gain control adjusted to give exactly five divisions deflection, for rise time measurements. The two traces were recorded separately, only one probe at a time being connected to the test waveform, Figure 3.31(b) being a double exposure.

With the timebase speed increased to 10 ns/div., the rise and fall times were measured as 4.5 and 4.0 ns respectively, implying a bandwidth, estimated by the usual formula, of around 80 MHz, even before making corrections for the rise times of the oscilloscope and test waveform. But there is a price to be paid for this performance, for the *CLC425* is a bipolar device with a typical input bias current of 12 μA. This means that the usual 10 MΩ input resistance is quite out of the question. In the circuit of Figure 3.31(a), however, a 100 kΩ input resistance has been arranged with the aid of an offset-cancelling control. In the sort of high speed circuitry for which this probe would be appropriate, an input resistance of 100 kΩ will often be acceptable. The need to adjust the offset from time to time is a minor drawback to pay for the high performance provided by such a simple circuit.

As described in connection with the unity gain active probe of Figure 3.29, the two ×10 versions of Figures 3.30(a) and 3.31(a) can be provided with clip-on capacitor caps for dc blocking. Clearly, with an active probe having a gain of ×10, the maximum permissible input signal, if overloading is to be avoided, is even lower than for a ×1 active probe. But it is not worth bothering to make a 20 dB attenuator cap for a ×10 active probe; with the probes described being so cheap and simple to produce, it is better simply to use a ×1 probe instead. An interesting possibility for the circuit of Figure 3.30(a) is to fit a miniature SPCO switch arranged to select either the 47 Ω resistor shown, or a 910 Ω resistor in its place, providing an active probe switchable between gains of ×1 and ×10. In the ×1 position, the bandwidth should rival or exceed that of Figure 3.29(a). This scheme is not applicable to the circuit of Figure 3.31(a), however, since while the *OPA655* is unity gain stable, the *CLC425* is only stable at a gain of ×10 or greater.

For a really wideband active probe

The three probes described so far all use opamps with closed loop feedback to define a gain of twice the net gain at the oscilloscope input. But another possibility is to use a unity gain buffer, where no external gain setting resistors are required. This provides the ultimate in circuit simplicity for an active probe. Devices such as the National Semiconductor FET-input buffers *LH0033* or *LH0063* could

138 *Analog circuits cookbook*

be considered. But having some samples of the MAXIM *MAX4005* buffer to hand, an active probe was made up using this device, which claims a 950 MHz –3 dB bandwidth and is designed to drive a 75 Ω load. The usual 10 MΩ probe input resistance is simply achieved, as the *MAX4005* is a FET-input device. The circuit is shown in Figure 3.32(a), it was made up on a slip of copper-clad laminate 1.5 cm wide by 4.0 cm long. The chip was mounted near one end, most of the length being taken up with arrangements to provide a firm anchorage for the 75 Ω coax. The chip was mounted upside down on four 10 nF chip decoupling capacitors connected to the supply pins and used also as mounting posts. Note that to minimise reflections on a cable, the *MAX4004* contains an internal thin-film output resistor

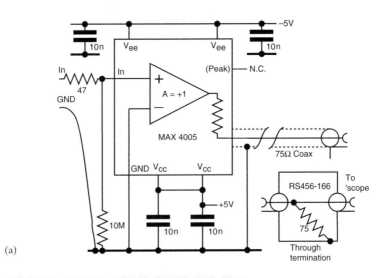

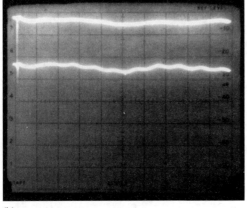

Figure 3.32 *(a) Circuit diagram of a wideband FET input probe with a gain of ×0.5. (b) Roughly level output of a sweeper used to test the probe circuit of (a) (upper trace) and output of probe (lower trace). Span 0–1000 MHz, IF bandwidth 1 MHz, 10 dB/div. vertical, ref. level (top of screen) +10 dBm*

to back-terminate the cable. This means in practice that the net gain from probe input to oscilloscope input is in fact ×0.5. This means in turn that the 5 and 10 mV input ranges on the oscilloscope become 10 and 20 mV respectively – no great problem – whilst, slightly less convenient, the 20 mV range becomes 40 mV/div. For this probe, of course, a 75 Ω coax lead was chosen, terminated at the oscilloscope input with a commercial 75 Ω through termination.

The expected bandwidth of this active probe being far in excess of the 250 MHz bandwidth of my *TEK 475A* oscilloscope, some other means of measuring it was required, and my *HP8558B* spectrum analyser was pressed into service. This instrument unfortunately does not provide a tracking generator output, but a buffered version of the swept first local oscillator output (covering 2.05–3.55 GHz) is made available at the front panel. In an add-on unit as described in Ref. 2, this is mixed with a fixed frequency 2.05 GHz oscillator to provide a swept output tracking the analyser input frequency. The mixer output is amplified and lowpass filtered, providing a swept output level to within ±1 dB or so, at least up to 1 GHz, at a level of around +6 dBm. This is shown as the top trace in Figure 3.32(b).

The active probe was then connected to the output of the sweep unit, via a 10 dB pad to avoid overloading, and a 50 Ω through termination to allow for the high input impedance of the *MAX4005*, taking great care over grounding arrangements at the probe input, Figure 3.33(b). The output of the probe (including the 75 Ω through termination shown in Figure 3.32(a)) was connected to the input of the spectrum analyser. This means that the 75 Ω coax was in fact terminated in 30 Ω. This mismatch explains the amplitude variations in the probe output, Figure 3.32(b), lower trace, corresponding to the electrical length of the 75 Ω coax lead. These apart, the level follows that of the sweeper output, upper trace, up to just under 1 GHz, where the expected roll-off starts to occur. The level is about 20 dB below that of the sweeper output which is explained by the 10 dB pad, and the additional loss above the expected 6 dB, due to the mismatch at the analyser input, see Figure 3.33(c).

The enquiring reader will have been asking 'What is the use of a 950 MHz bandwidth active probe when the 75 Ω termination at the oscilloscope is in parallel with an input capacitance of around 20 pF?' After all, the effective source resistance seen by the 'scope input is 37.5 Ω (the 'scope bridges both the source and load resistors, which are thus effectively in parallel) while the reactance of 20 pF at 950 MHz is 8.4 Ω. But it must be remembered that the figure of 20 pF is a lumped figure, measured at a comparatively low frequency. In fact, this capacitance is typically distributed over a length of several inches, the input attenuator in the 475A, for example, being

(a)
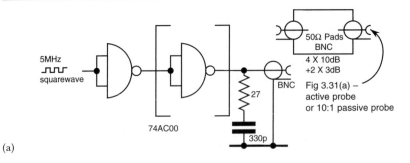

(b)
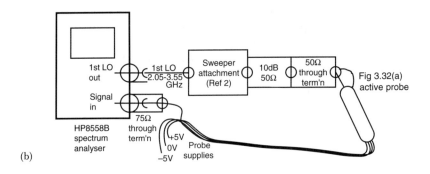

(c)
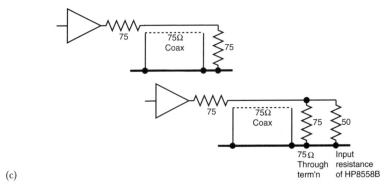

Figure 3.33 *(a) Test circuit used to produce a 5 MHz squarewave with fast edges, to test the probe of Figure 3.31. The 27 Ω plus 330 pF snubber at the output suppressed ringing on the test waveform. (b) Test set-up used to test the wideband probe of Figure 3.32. (c) Showing how the 6 dB signal reduction in normal use becomes 11 dB in the test set-up of (b) above. Together with the 10 dB pad at the sweeper output, this accounts for the 21 dB separation of the traces in Figure 3.32(b)*

implemented in thick film pads. These are connected in circuit or bypassed as required by a series of cams on the volts per division switch. Thus the 20 pF is distributed over some kind of transmission line, the characteristics of which are not published. It is therefore likely that the effective capacitance at 950 MHz is less than 20 pF: the only way to be really sure what bandwidth the probe of Figure 3.32(a) provides with any given oscilloscope is to measure it. But given the 370 ps rise time of the *MAX4005*, this exceedingly simple active probe designed around it is likely to outperform the vast majority of oscilloscopes with which it may be used.

References

1. Dearden, J. (1983) 500 MHz high impedance probe. *New Electronics*, 22 March, p. 28.
2. March, I. (1994) Simple tracking generator for spectrum analyser. *Electronic Product Design*, July, p. 17.

Acknowledgements

Figures 3.25–3.28 are reproduced from Hickman, I. (1995) *Oscilloscopes: How to Use Them How They Work* – 4th edn, ISBN 0 7506 2282 2, with the permission of the publishers Butterworth-Heinemann Ltd.

4 Measurements (rf)

> **Amplitude measurements on rf signals**
>
> Amplitude measurements on rf signals require a detector of some sort. Many types exist and the following two articles examine the performance of some of them.

Measuring detectors (Part 1)

A detector of some sort is required in order to measure the amplitude of an ac signal. In the case of an amplitude modulated carrier, e.g. a radio wave, measuring its amplitude on a continuous basis will extract the information which it carries. One of the earliest detectors was the coherer, a glass tube filled with iron filings which, when an rf current passed through them, tended to stick together. This reduced the resistance in the circuit containing a local battery, causing it to operate the tape-marking pen of a morse inker. (A tapper was also needed to re-randomise the filings after the received dot or dash, to re-establish the initial high resistance state.)

I have never used one of these primitive but intriguing devices, but I early gained some practical experience of a later development, the crystal detector. This permitted the demodulation of amplitude modulated waves carrying speech or music, something beyond the capability of the coherer. The crystal detector – usually a lump of galena, an ore of lead – held sway for some years, but by sometime around the mid-1930s the standard domestic receiver was a superhet mains table ratio. The detector circuit generally looked something like Figure 4.1(a), where the same diode (thermionic of course) is shown used for both demodulation and to produce a voltage for automatic gain control (AGC), a common arrangement – although

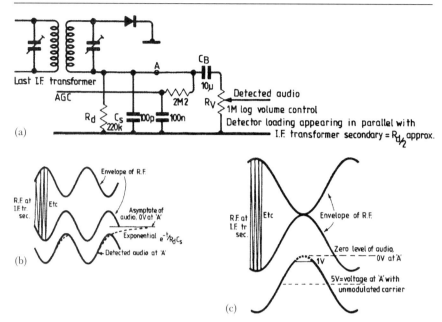

Figure 4.1 *The simple diode detector as fitted to AM broadcast receivers introduces high levels of distortion because the AF filter components prevent the detector from following the rf envelope*

often a second diode section of a double-diode-triode was used for the latter function. This deceptively simple circuit is not a particularly 'good' arrangement, being fraught with various design compromises, the unravelling of which is an instructive and (I hope) interesting exercise in practical circuit design.

The first concerns the time constant $C_s R_d$ formed by the detector load resistor and the rf smoothing capacitor. Demodulation of the peaks of the rf envelope presents the circuit with no particular problems. However, with the typical values shown, $C_s R_d$ has a 3 dB corner frequency of 8 kHz; not much above the highest frequency components of 4.5 kHz found in medium and much long wave broadcasting. Consequently, in the case of a large amplitude signal at a high audio frequency such as 4.5 kHz, the detected output could come 'unstuck' on the troughs of modulation, R_d being unable to discharge C_s rapidly enough (Figure 4.1(b)), resulting in second harmonic and higher even-order distortion products. One could of course reduce C_s, but there are only six and a bit octaves between 4.5 kHz and the intermediate frequency of around 465 kHz (still in common use) in which to achieve adequate suppression of the rf ripple.

A further subtle problem centres on the blocking capacitor C_b and the volume control R_v. The dc load on the detector is 220 kΩ but at ac the 1 MΩ resistance of the volume control appears in parallel with it as well. C_b will be charged up to the peak level of the unmodulated carrier, say –5 V at its junction with C_s, and being large in order to pass the lowest notes, it will simply appear in the short term as a 5 V battery. At the trough of, say, 100% modulation, +4 V will appear across the volume control whilst –1 V appears across R_d. Thus the circuit can only cope with a maximum of 80% modulation and, C_b being large, this limitation applies equally at all audio frequencies. In fact, the situation is rather worse than this, as the AGC line contributes another ac coupled load, further reducing the ac/dc load ratio and thus compounding the even-order harmonic distortion which results.

A circuit very similar to Figure 4.1(a) but with different component values, e.g. a 4k7 volume control, was used in transistor portables implemented with discrete PNP transistors, with similar problems. Thus the simple diode detector is adequate for domestic entertainment purposes, but some improvements are needed if it is to be used as the basis of a measuring instrument. Indeed, the basic diode detector circuit is so poor that, at frequencies where alternative circuits employing opamps are feasible, they are usually nowadays preferred.

The advantage of the diode detector is that it can be used at much higher frequencies, fairly successfully where suitable circuit enhancements are used to avoid some of its limitations. One of the most serious of these is its restricted dynamic range. As its cathode (when used to provide a positive output) is connected to an rf bypass capacitor across which the peak value of the rf signal is stored, obviously the peak-to-peak rf input voltage must be restricted to less than the diode's reverse voltage rating. This sets an upper limit to the dynamic range, though where the detector is preceded by an amplifier, the output swing available may in practice be the limiting factor. But not always; some Schottky diodes suitable as UHF detectors have a maximum reverse voltage rating of only 5 V or even less.

For large inputs, the relation between the detected dc output amplitude and the amplitude of the ac input is linear, that is to say that equal increments in the ac input result in equal increments in the detected voltage. However, this is not to say that the detected voltage is strictly proportional to the ac input: in fact isn't quite. The detected output is less than the peak value of the ac input voltage by an amount roughly equal to the diode's 'forward drop'. So the relation, though linear at high levels, is not proportional; projected backwards as in Figure 4.2(a) it does not pass through the origin. The characteristic looks, indeed, very like the static dc characteristics of the diode. The non-linear portion at the bottom of the curve exhibits a square-law characteristic, so that at very low input levels indeed, doubling the ac input results in four times the detected dc output. The diode can still be used in this region, provided due allowance for the changed characteristic is made. In fact the only limit on how small an ac signal the diode can be used to detect is that set by noise: obviously the less noisy the diode, the more sensitive the equipment employing it, e.g. a simple diode/video radar receiver.

A convenient practical method of measuring a diode's noise-limited sensitivity uses a signal generator with a pulse modulation capability. Squarewave on/off modulation is used, and the resultant detected output is displayed on an oscilloscope, as in Figure 4.2(b). The carrier level which just results is no overlap of the 'grass', but in no clear space between the two levels of noise either, is known as the 'tangential sensitivity'. This is not an exact measurement, since the measured level will depend to some extent upon the oscilloscope's intensity setting, but in practice the variation found when a given diode is measured on different scopes by different people is not large, and since it is so simple to carry out, the method is popular and widely used.

In some applications, a diode detector may be used in the square-law region without any linearisation, or with some approximate

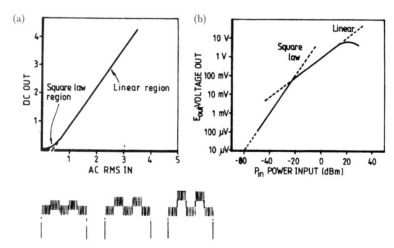

Figure 4.2 The detection efficiency of the diode detector falls sharply at low signal levels within the square-law region of the diode's response, as the two graphics show. A squarewave modulated carrier can be used to determine sensitivity (see text)

linearisation over a limited range, using an inverse square-law circuit. This can provide useful qualitative information, as in the diode-video receiver already mentioned. But to obtain quantitative information, i.e. to use a diode as a measuring detector down in the square-law region, some more accurate means of linearising the characteristic is needed. Nowadays, what could be simpler than to amplify the resultant dc output with a virtually drift- and offset-free opamp, for example a chopper type, pop it into an ADC (analog to digital converter) and use some simple DSP (digital signal processing) on the result? This could take into account the output of a temperature sensor mounted in the same head as the detector, together with calibration data for the characteristic of the particular diode fitted. However, an alternative, venerable and very elegant scheme is shown in Figure 4.3. Here, two matched diodes are employed, fitted close together in the measuring head, but screened from each other and kept at the same temperature by the surrounding metal work. A differential amplifier compares the output of the two diodes and controls an attenuator situated between an oscillator and a level indicator. The former works at a convenient comparatively low frequency and high level, so that high linearity is

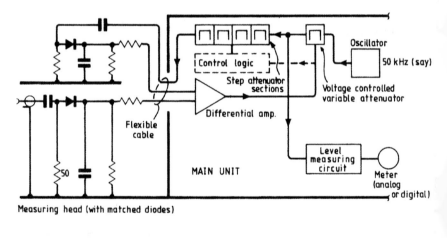

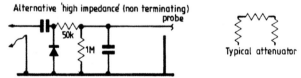

Figure 4.3 *Accurate rf level measurement requires linearisation of the diode response. This can be done by using a dummy diode and separate rf reference source for comparison with the rectified level of the signal under test*

easily achieved in the latter. Further attenuation can be introduced in steps, to allow for ranges down to the tangential sensitivity, either manually by the operator (in which case the need for a range change is indicated by the meter's reading above or below the calibrated part of the scale) or automatically. Provided the loop gain is high, the stability of the output level of the oscillator is not critical, the accuracy of the measurement depending only upon that of the level meter, the step attenuators and, of course, the matching of the diodes.

Useful though this scheme is in an rf millivoltmeter working up to a few GHz, it is mainly used for static level measurements, as clearly the speed of response is limited. Where a faster response, covering a large dynamic range is required, other schemes, no less ingenious, can be used.

Measuring detectors (Part 2)

The useful dynamic range of a diode detector can be extended by applying a small amount of dc forward bias. There is also the standing offset (temperature dependent) to cope with, but that can be balanced by another dummy diode circuit, as in Figure 4.4(a). The forward bias has another benefit: when the input signal falls rapidly the detected output voltage falls aiming at the negative rail, rather than 0 V as with the diode detector in Figure 4.1. If the negative rail voltage is large, R virtually represents a constant current 'long tail', defining a negative-going slew rate limit for the detector of $dv/dt = (V-)/CR$. In this case, if the detected output parts company with a trough of the modulation, it will not be towards the tip as in Figure 4.1, but at the point of maximum slope. For sinewave modulation of $v = E_{max} \sin(\omega t)$, this will be given by dv/dt, which equals $E_{max} \omega \cos(\omega t)$. The maximum value of $\cos(\omega t)$, of course, is just unity and occurs when $\sin(\omega t)$ equals zero, so $dv/dt_{max} = (\omega E_{max})$ volts per second, giving the minimum permissible value for $(V-)/CR$ for distortionless demodulation.

From Figure 4.4(a) it is but a small step to replace the detector diode with a transistor, giving an arrangement which in the days of valves was known as the infinite impedance detector (Figure 4.4(b)). With no rf voltage swing at either anode or cathode, a triode was perfectly satisfactory and, assuming no grid current, the only loading on the preceding tuned circuit was the loss component of the $C_{grid\text{-}all}$ capacitance. This was very low up to VHF and quite negligible at all the usual intermediate frequencies then in use. In the case of Figure 4.4(b), clearly the loading is finite, however low the frequency, but it

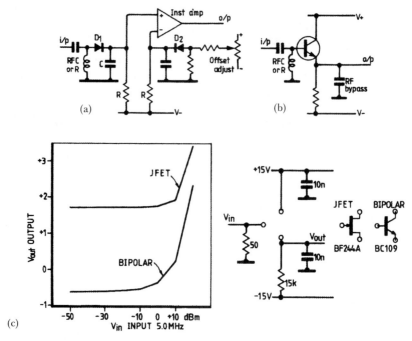

Figure 4.4 (a) DC bias to the diode improves linearity by several dB. If R is made high enough, it becomes a current source greatly extending the linear detection region but this also requires a larger negative rail voltage. (b) Functional equivalent of the diode circuit (a). (c) Comparing the performance of a JFET versus bipolar infinite impedance detector. The latter has a more abrupt cut-off providing a higher dynamic range

will be less than for the diode of Figure 4.4(a) by a factor roughly equal to the current gain of the transistor. Substituting an rf JFET such as a *BF244* results in a very close semiconductor analogy of the infinite impedence detector. In either case, a balancing device may be added as in Figure 4.4(a) if the absolute detected dc level is important. Figure 4.4(c) compares the performance of a JFET and a bipolar infinite impedance detector; as is to be expected, the more abrupt cut-off of the latter (higher g_m) results in a higher dynamic range.

The circuit of Figure 4.4(b) lends itself to a further improvement not possible with the simple diode circuit Figure 4.5(a). Here, the collector current of Tr_1 in the absence of any input signal is arranged to be much smaller than the current through R_3, which is thus mainly supplied via Tr_2. When a large input signal is applied, once the steady state condition has established itself, Tr_1 conducts only at the tips of positive-going half cycles. These current pulses are amplified by Tr_2, increasing the tail current through R_3, thereby holding Tr_1 cut off

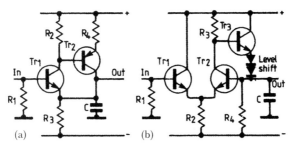

Figure 4.5 (a), (b) Active detectors provide further improvements on the infinite impedance detectors

except at the very tip of each cycle. The input impedance may not be quite as high as the infinite impedance detector and is also slightly non-linear, due to the voltage swing across R_2 appearing across the collector base capacitance C_{cb} of Tr_1. At low input levels, Tr_1 never cuts off but passes a distorted sinewave where the increase in current on positive swings of the input is greater than the decrease on negative swings. Tr_2 never cuts off either, so the voltage swing at its base is very small and there is little Miller feedback via Tr_1's C_{cb}. Tr_2's collector current is modulated, increasing more on the positive swings of the input and decreasing less on negative swings, so increasing the average voltage at Tr_1's emitter. The circuit is in effect a servo-loop or NFB system, which is linear as far as the envelope of the rf input is concerned, but non-linear over each individual cycle of rf. Tests on the circuit showed a linear dynamic range approaching 60 dB, measured in the upper part of the HF band.

Figure 4.5(b) shows another variant, with some rather nice features. The inverting PNP stage of Figure 4.5(a) has been replaced by an emitter follower; an inversion is not required with this circuit as Tr_1 base to Tr_2 collector is non-inverting. There is now no rf voltage at Tr_1's collector at any input level, and the input impedance should be as high as the infinite impedance detector. Although the circuit uses more components than Figure 4.5(a), in an integrated circuit implementation this is of little consequence.

The circuits shown in Figures 4.4 and 4.5 measure the amplitude of the positive peak of the input signal, and this will be a good guide to its rms value if the input is taken from a tuned circuit, and so virtually undistorted. In the case of a wideband detector, however, the wanted input signal may be significantly distorted and this may affect the expected 1.414:1 ratio of peak to rms voltage. I say 'may' because in the case of both odd-order and even-order distortion, the measured peak voltage could in fact be the same as if the distortion components (harmonics) were just not there. More commonly though, the peak

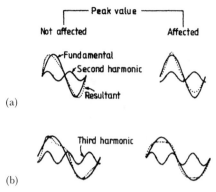

Figure 4.6 In a wideband detector, measuring the input signal's positive peak may affect the expected ratio of peak-to-rms voltage. (a), (b) show resultant phases in second and third harmonics

voltage will be affected (Figure 4.6). An even-order component, e.g. second harmonic, will reduce the amplitude of one peak but increase the amplitude of the opposite polarity peak by the same amount. It follows that by measuring the amplitude of both peaks and taking the difference – i.e. using a peak-to-peak detector – no error results, and the rms value of the fundamental component, if that is what you want to measure, is just the peak-to-peak value divided by 2.828. A difference between the absolute values (moduli) of the positive and negative peaks not only indicates the presence of distortion, but also directly gives the value of the sum of the in-phase components of even-order distortion present. Odd-order components, e.g. third harmonic, affect both peaks in the same way: not only will they alter the expected 1:414:1 peak-to-rms ratio, but unlike even-order components there is no convenient indication (such as unequal +ve and –ve peaks) of their presence.

An alternative to measuring peak values or peak-to-peak values is to measure the average value of the modulus of the input sinewave – the average value of a sinewave itself is of course zero. This takes us to the topic of ideal rectifiers, which are readily implemented with opamps. Such circuits are limited to audio and video or low rf frequencies, but Figure 4.7 shows a circuit which is average responding, linear down to very low levels and will work up to VHF with suitable components. Twenty years ago I designed it into low-level measuring sets operating up to 20 MHz, for supply to the GPO. It operates as a product detector, where the amplified signal is used to provide its own switching (reference) drive. In principle it operates linearly down to the point where there is no longer enough drive to the four-transistor switching cell. In practice, the limit may be where the differential output signal reverses sense, due to device offsets. For use up to VHF, it may be necessary to introduce delay into the signal path to compensate for the lag through the switching drive amplifier, as shown in Figure 4.7.

A little simple algebra shows that the average value of a sinewave is related to the rms value by $E_{av} \times \pi/2 \times 0.707 = E_{rms} = 1.11 E_{av}$. The

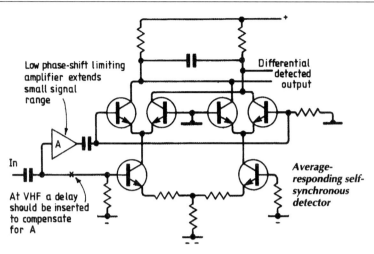

Figure 4.7 *Circuit which is average responding, linear down to very low levels and will work up to VHF with suitable components*

presence of even-order harmonics does not affect the measured value of the fundamental, but the same is not true of odd harmonics. However, whereas 10% of third harmonic will give an error in a peak reading somewhere between 0 and 10%, for an average-responding detector, the error is between zero and only 3.3%, i.e. one-third of the harmonic amplitude. For the fifth harmonic, the maximum possible error is only one-fifth and so on for higher odd harmonics. So an average-responding circuit is really quite useful.

An LCQ test set

The instrument described here enables the values of inductors and capacitors to be measured at or near the frequency at which it is intended to operate them, up to around 150 MHz. In the case of inductors particularly, the results may be quite different from a measurement made at audio frequency on an ordinary *LCR* bridge. It also permits estimates of inductor Q at the working frequency.

Measuring *L* and *C* at frequency – on a budget

In the development labs of large companies, measurement of inductance or capacitance is very simple. One simply connects the component to be measured to a network analyser and makes an s_{11}

measurement. Using the marker function, a screen readout of the capacitance (or inductance) and the associated loss resistance (or the real and imaginary part of the impedance) at the frequency of interest is obtained. The change of apparent value with frequency can also be displayed on a Smith chart presentation. Unfortunately, on returning to his home laboratory, the typical electronics engineer interested enough in the subject to pursue it away from work, finds himself bereft of such aids. The price of a network analyser, for example, is around £15 000. Provided one only requires to measure capacitors, there is no great problem since digital capacitance meters are cheap and readily available at less than £50, whilst many designs for constructing one's own have appeared over the years. Most capacitors are near-ideal components, so the frequency at which they are measured is largely immaterial – unless that is you wish to know just what the loss resistance is at a given frequency, in which case you will need a much more sophisticated (and expensive) measuring instrument. With inductors, the measurement problems are much more severe, since an inductor is really only usable over about two decades of frequency, at least for air-cored types. At higher frequencies, the inductor resonates with its own self capacitance, whilst at about a hundredth of that frequency, its Q has dropped to the point where it is of little use in a practical circuit. There is thus a niche for a cheap-to-build instrument which will measure capacitors and, more particularly, inductors at, or close to, the frequency at which it is intended to use them. Such a device is described below.

Frequency choices

The traditional method of measuring inductance and capacitance is the Q meter. Models were available from manufacturers such as Hewlett Packard, Advance, Boonton and Marconi. From the last mentioned, a well-known early model came in a box almost a foot deep, with all controls, meters, etc. on the 'front' panel, namely the top surface which was about two feet square. The highest operating frequency for this model was 25 MHz and, perhaps for this reason, decade multiples of 250 kHz are common frequencies for Q measurements. The other common frequencies are decade multiples of 790 kHz. This may be for one or both of two reasons: firstly, it is roughly $\sqrt{10}$ times 250 kHz, giving two (geometrically) equally spaced spot test frequencies per decade; and, secondly, it is half of the frequency corresponding to 10^7 radians per second. Anyway, since the Q of commercially available inductors, such as those used in this design, is commonly quoted at these frequencies, they were selected for the internal test frequency generator in the following design.

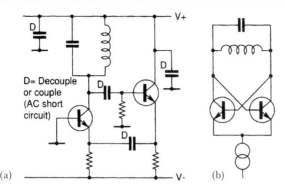

Figure 4.8 (a) Two transistor oscillator looks at first sight like an emitter follower driving a grounded base stage. But the earth point is an arbitrary convention. (b) If the decoupling capacitors in (a) are shown as short-circuits at rf, the circuit is seen to be a balanced push–pull oscillator

The basic circuit of the test generator, shown in Figure 4.8(a), is seen to be a two-transistor circuit, which looks at first sight like an emitter follower driving a grounded base stage, and can indeed be analysed as such. But in fact it is functionally equivalent to the push–pull oscillator of Figure 4.8(b). In any half cycle of the voltage appearing across the tank circuit, one transistor is cut off whilst current through both of the tail resistors flows through the other transistor. Thus the tank circuit receives the total tail current, chopped up into a (near) squarewave. The transistors act largely as switches and the amplitude of the tank voltage is given by its dynamic resistance R_d times the fundamental component of the current squarewave.

Circuit details

The full circuit of the test set (excluding power supplies) is given in Figure 4.9, which shows that tank circuits giving seven fixed spot test frequencies are available, together with a facility for feeding in an external test signal of any desired frequency. The same LC ratio is employed for all the tank circuits, so that they all have the same R_d (about 4k0), or would do if the Qs were all equal, which is roughly the case. This figure is reduced to about 700R by the shunting effect of $(R_1 + R_2)$, 470R and 5K6, and R_6, 1K giving a loop gain from Tr_2 base to Tr_1 collector of roughly 700 divided by R_4, about ×14 or well in excess of unity, ensuring reliable oscillation. Given a total tail current via R_3 and R_5 of around 10 mA, this provides a large enough swing across the tank circuit to chop the tail current into a respectable squarewave, ensuring the amplitude of oscillation varies little from range to range.

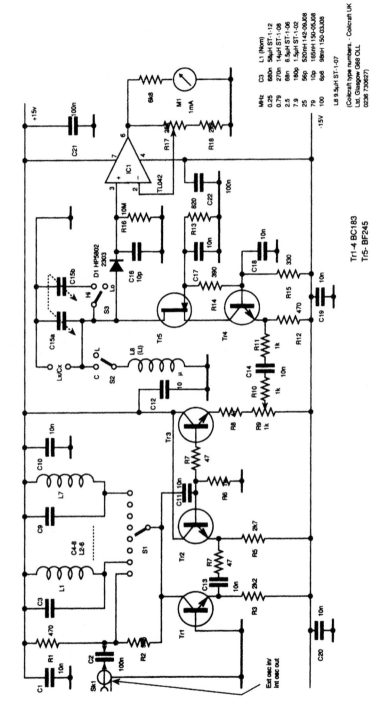

Figure 4.9 *The switched frequency rf source (Tr_1 to Tr_3) provides a constant level drive source to the reactance under test, and detector/measurement circuit D_1, IC_1. NOTE: R_1 is 47Ω, 47R resistor at C_{13} is R_4, emitter follower at R_8 is TR_3*

On the other hand, in the EXTernal OSCillator IN position of S_1, the collector load of Tr_1 is reduced to 47R, giving a loop gain of less than unity and thus preventing oscillation. The RF tank voltage (or EXT OSC input) is buffered by Tr_3, the output of which drives a test current (determined by the setting of R_9) into a cascode composed of Tr_4 and Tr_5. The output admittance of a cascode stage is very low – especially when the first transistor is driven in grounded base – so that the test circuit is driven by a near ideal constant current generator. Of course, at the higher frequencies, the cascode's output impedance will fall, but so will the R_d of any practical circuit that you are likely to want to measure. This arrangement is thus adequate for the purpose, and much easier to implement than the traditional Q meter scheme, where an RF current (measured by a thermocouple meter) was passed through a very low resistance placed in series with the LC circuit under test.

The voltage across the inductance under test, resonated with C_{15}, is detected by D_1, which places very little loading on the circuit owing to the high value of the following dc load, R_{16}. IC_1 acts as a buffer to drive the meter M_1. The gain of the buffer stage is adjustable over the range unity to ×12 by means of R_{17}. The tuning capacitor C_{15} is a 500 pF twin gang type, where one half has had all the moving plates except one removed. This reduces the maximum capacitance of C_{15a} to around 45 pF, including the stray capacitances added by S_2, S_3, D_1 and Tr_5. For use at lower frequencies, S_3 switches the 500 pF section in parallel, enabling a wide range of inductors to resonate over the range 250 kHz to 79 MHz, or even 100 MHz (using tank circuit $C_9 L_7$).

For measuring capacitors, the test inductor $L_t(L_8)$ is switched into circuit, and resonated with C_{15} near maximum capacitance. The unknown capacitor is then connected to the test terminals (an Oxley pin projecting through the panel and an earth tag), and resonance restored by reducing C_{15}. The change in C_{15} capacitance gives the effective capacitance of the unknown capacitor at the test frequency used.

Constructional tips

This simple test instrument has proved very useful, but naturally for best results some care is needed both in construction and use.

The prototype was constructed in a diecast box, to guarantee the absence of direct coupling between an inductor under test and whichever tank circuit was in use. A compact construction, especially around the test terminals, is essential to minimise stray inductance and capacitance which could cause problems at 100 MHz. To achieve this end while keeping the mechanics simple, the capacitance scales

156 *Analog circuits cookbook*

have been placed on the side of the box, while all other controls are on the top (Figure 4.10). Miniature or, better, subminiature components are recommended, especially for S_2 and S_3. The use of a ground plane is recommended: in the prototype this was simply a sheet of single sided copper clad SRBP which was clamped to the underside of the front panel by the mounting bushes of S_1 and R_9 and connected by a wide piece of copper tape to the frame of C_{15}. Fresh air construction was used for all those parts of the circuitry operating at rf, 10 nF decoupling capacitors being soldered to the ground plane wherever needed. Their other ends were used as mounting points for the other components, a form of construction which is crude and ugly as it is cheap and effective. As there was no intention to put the unit into production, there was no point in going through iterations to optimise PC layout. IC_1 was mounted on a scrap of strip board soldered to the groundplane, with the supplies brought in from the power unit mounted in the base of the box via a plug and socket.

Calibration

Calibration presents some interesting problems, which can be solved with the aid of four or five 100 pF 1% capacitors. Using various series/parallel combinations of these, one can make up capacitances of 20, 25, 33, 50, 67, 100, 125, etc. up to 500 pF. However, the problem is how to take into account the stray capacitance associated with the test circuit. (If the unit were only going to be used for measuring capacitors, the internal stray capacitance could be ignored, and the scale simply calibrated in terms of the capacitance added at the test terminals. But to measure inductors, knowing the frequency at which the circuit is

Figure 4.10 *Completed test set, showing controls on two faces of the box*

resonated, requires also a knowledge of the 'true' total circuit capacitance.) The first step is to assemble the unit and fit the pointer knob of C_{15}. Now, with the capacitor fully in mesh, make a fiducial (reference) mark on the blank scale, so that the knob can always be refitted in exactly the same position if subsequently removed. Set S_2 to 'C', S_3 to LO, the gang to minimum capacitance and connect a capacitance of 25 pF to the test terminals. Feed in an external test signal, and note the frequency at which resonance is indicated. Now increase the capacitance to 33 pF and repeat the procedure. From these results, the method shown in 'Quantifying internal capacitance', later in this section, will give a close approximation to the test circuit's true internal capacitance. Knowing this, the various combinations of the 100 pF capacitors can be used to calibrate the HI and LO scales, making due allowance for the internal capacitance. The spot test frequencies of 250, 790 kHz, 2.5, 7.9, 25, 79 and 100 MHz should now be set up, by adjusting the cores of L_1 to L_9 respectively. For this purpose, the frequency can be monitored at BNC coaxial socket SK_1.

Operation

In use, R_{17} should normally be kept set anticlockwise, at the minimum gain setting, with just enough drive applied to the test circuit by R_9 to give full scale deflection. Under these conditions, the rf signal into the detector is large enough to give a linear response. So, by detuning either side of resonance to 71% of meter FSD and noting the two capacitance values, the Q of the inductor under test can be estimated. (If the average of the two values, divided by their difference, is 25, then the Q is 25, courtesy of an approximation based upon the binomial theorem for values of $Q>10$.)

At higher frequencies, where the lower value of the R_d of the test circuit is such that full scale deflection cannot be achieved even with R_9 at maximum, R_{17} should be advanced as necessary. As mentioned earlier, capacitors are measured by switching the test inductor L_t (L_8) into circuit and noting the reduction in the value of C_{15} required to restore resonance when the unknown capacitor is connected to the test terminals. As the Q of the capacitor under test is likely to be greater than that of L_t, estimation of the capacitor's Q is usually not possible – a limitation the instrument shares with traditional Q meters.

Higher spot frequency

Incorporating a higher spot test frequency, say 250 MHz, is not possible with the transistors used – even 144 MHz proved unattainable. Using higher frequency transistors should in principle provide the answer, but it is then very difficult to avoid parasitic oscillations due to stray inductance and capacitance associated with S_1. A really miniature S_1 might do the trick, but a better scheme would be a separate 250 MHz oscillator and buffer, powered up when Tr_{1-3} were not and vice versa. However, although of course it lacks the convenience in use of a network analyser, even as it stands, the instrument is a great advance upon nothing at all. Even without using an EXT OSC, it is always possible, using the nearest spot frequency, to measure an inductor at a factor of not more than the fourth root of ten removed from the intended operating frequency, over the range 140 kHz to 178 MHz.

Quantifying internal capacitance

$$\omega_1 = \frac{1}{(LC)^{0.5}} \qquad \omega_2 = \frac{1}{(L(C+C_1))^{0.5}}$$

$$\frac{\omega_1^2}{\omega_2^2} = \frac{L(C+C_1)}{LC}$$

If C_1 is known then C is determined. Let:

$$\left(\frac{\omega_1}{\omega_2}\right)^2 = 1+\Delta$$

then

$$1+\Delta = \frac{C+C_1}{C} = 1+\frac{C_1}{C}$$

So

$$\Delta = \frac{C_1}{C} \qquad C = \frac{C_1}{\Delta}$$

For example, if $C = (25p + C_{stray})$ and $C_1 = 8.33 \, (= 33.3p - 25p)$ then:

$$C_{stray} = (C - 25p) = \left(\frac{C_1}{\Delta}\right) - 25p$$

$$= \left(\frac{8.33}{\Delta} - 25\right)_p$$

This assumes F_1 is well below the self-resonant frequency of L, so that L is effectively the same at F_1 and F_2.

Equivalent circuits of inductors and capacitors

In addition to series loss component r_s and a series inductance L_s, a capacitor has a shunt loss component R_p. Except in the case of electrolytic capacitors, R_p is usually so high that it may be ignored. At the frequency F_r where C resonates with L_s, the capacitor looks resistive, and looks inductive above this frequency. For a tantalum electrolytic of a few microfarads, F_r is usually around 100 kHz, with a very low Q. Dissipation due to the loss resistance r_s determines the maximum current that a capacitor, e.g. a mica type in an RF PA, can safely carry. Care should be taken when paralleling two decoupling capacitors, since for some types the Q at series resonance can be quite high. If of the same (nominal) value, one may resonate at a somewhat lower frequency than the other: at a slightly higher frequency its inductive reactance can be parallel resonant with the other capacitor – result, no decoupling at that frequency!

Good practice is to make one capacitor at least ten times as large as the other.

Where a capacitor has a parasitic series inductance L_s, an inductor has a parasitic shunt capacitance C_p. This cannot accurately be considered lumped, being distributed between the various turns of the winding. Inductors are much more imperfect components than capacitors. Whereas the latter can be used over a frequency range of 10^7:1 or more, the range between the self-resonant frequency of an inductor and the frequency at which its Q has fallen to an embarrassingly low value is as little as 100:1 – at least for air-cored types, including those with a slug adjuster. High A_l inductor pot cores can provide a large inductance with very few turns, reducing C_p, especially if the turns are spaced, resulting in a wider useful operating frequency range.

Inductor   Capacitor

A spectrum monitor

Encounters with rf are much easier if a spectrum analyser is to hand. Although based on a commercial TV tuning head, this design delivers linear, useful performance in its basic form and may be adapted to a much higher degree of sophistication including continuous coverage and wider frequency span.

Add on a spectrum analyser

Introduction

In general electronic design and development work, fault-finding, servicing, etc. in either analog or digital areas, an oscilloscope is undoubtedly the basic tool of the trade. When investigating the performance of rf equipment, however, whilst an oscilloscope (with sufficient bandwidth) is a great help and certainly much better than nothing, it is very revealing to have a spectrum analyser to do the job. Unfortunately, a professional-standard spectrum analyser is very expensive; even a second-hand model will cost around £2000. A much cheaper alternative, which is capable of considerable further development, is described below.

As it stands, it has its limitations, so it should be thought of as a spectrum monitor rather than a spectrum analyser, to distinguish it from the real thing. Nevertheless, it has already proved itself extremely useful and would be even more so if the suggested lines for further development were pursued. Most serious electronics enthusiasts will, like the writer, already possess an oscilloscope. The monitor was therefore designed to use an existing oscilloscope as the display, a very basic oscilloscope being perfectly adequate for the purpose.

Circuit design

The spectrum monitor is built around a TV tuner, the particular one used by the writer being a beautifully crafted all surface-mount example, the *EG522F* by Toshiba, of which four were bought some years ago at a mere £5 each, along with half a dozen even cheaper (though rather untidily built) tuners of Italian manufacture. Whether the particular Toshiba model is still available is open to question, but a wide range of TV tuners is held in stock by various suppliers (see Hickman, 1992a). Such tuners offer a high degree of functionality at a price which is no higher than many an IC, so that they should be regarded simply as components.

The *EG522F* provides continuous coverage from the bottom of Band I to the top of Band III in two ranges, a third range covering Bands IV/V. It is a pity about the gap between the top of Band III and the bottom of Band IV, but when I enquired of the source there quoted the continuous coverage tuner mentioned in Hickman (1992b) was no longer available. However, note that continuous coverage is provided by tuners designed for VHF/UHF/Cable/Hyperband applications. The design of this spectrum monitor is generally applicable to most types of TV tuner and the reader may employ whatever tuner is to hand or can be obtained, any necessary circuit modifications being straightforward.

It was desired to give the finished unit as much as possible of the feel of a real spectrum analyser, albeit of the style in use ten or fifteen years ago, rather than the faceless all push-button controlled variety currently in vogue. The design challenge was to achieve this without introducing excessive complication, but to leave the way open for further development if required. To this end, within its case the monitor was constructed as three separate units – PSUs, sweep generator, RF/IF unit – interconnected by ribbon cables long enough to permit the units to be worked on whilst operating, out of the case. After some initial experimentation with the tuner and a sawtooth generator, design work started in earnest with the construction of a suite of stabilised power supplies. These were ±15 V for general analog circuitry, +12 V for the tuner and +30 V for its tuning varactor supply (Figure 4.11), terminating in a 7 pin plug accepting a mating ribbon-cable-mounted socket (RS 'inter PCB crimp' style). This done, the sweep-circuitry to drive the tuner's varactor tuning input was addressed in more detail and the circuit shown, in basic form, in Figure 4.12(a) was developed. This produces a sawtooth waveform of adjustable amplitude and fixed duration, the amplitude being always symmetrically disposed about ground. This means that as the 'span' (the tuning range covered by the monitor) is increased or decreased (the 'dispersion' decreased or increased), a signal at or

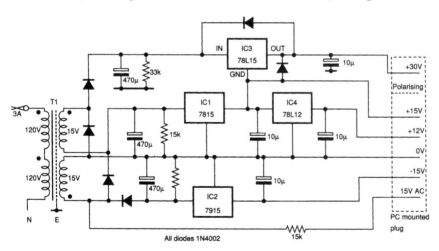

Figure 4.11 *Stabilised power supplies. Nominal 15 V secondaries produce 22 V dc raw supplies. The 15 V ac output provides a timebase for the sweep voltage generator*

162 Analog circuits cookbook

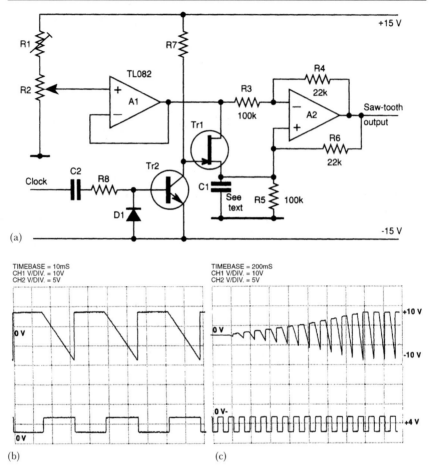

Figure 4.12 (a) Basic circuit of the tuning sweep generator, employing a Howland current pump (A₂ and associated resistors). (b) Output waveform shown in relation to the controlling clock waveform. (c) Advancing R₂ from ground to maximum increases the sweep width whilst remaining ground-centred

near the centre of the display becomes contracted or expanded widthwise but remains on-screen – a great convenience in use.

Operation is as follows. On negative excursions of the clock drive, Tr_2 is off and Tr_1 clamps the capacitor C_1 and the NI input of A_2 to the voltage at the output of $A1$, V_{clamp}: the output therefore also sits at V_{clamp}, the voltage at the wiper of R_2. A_2 forms a Howland current pump, so that when Tr_2 is turned on, removing the clamp, a negative charging current V_{clamp}/R_5 is applied to the capacitor. As A_2 must act to maintain voltage equality between its inputs, a linear negative going ramp results. If C_1 is selected correctly relative to the clock frequency, the voltage across it will just reach $-V_{clamp}$ during each positive excursion of the clock (Figure 4.12(b)). For convenience, the clock frequency is derived from the mains, giving a choice of sweep durations. The sweep amplitude can be set to any value from zero to maximum, the sweep remaining ground centred as illustrated in Figure 4.12(c), where R_2 was used to advance V_{clamp} steadily from ground to its maximum value, over a number of sweeps.

Figure 4.13 shows the full circuit of the sweep circuitry which operates as follows. The 15 V ac from the PSU is sliced by Tr_1 (Figure 4.13(a)) and fed to a hex inverter to sharpen up the edges. R_8 and R_9 around the first two inverters provide some hysteresis – without this, noise on the mains waveform will simply be squared up and fed to the counters as glitches, causing miscounting. The output of the inverters is a clean 50 Hz squarewave and appears at position 1 of switch S_{1B}. The half period is 10 ms, this setting the shortest sweep duration. A string of four *74LS90* decade counters provide alternative sweep durations up to 100 seconds. The selected squarewave (at nominal 5 V TTL levels) from S_{1B} is level shifted by Tr_2 and Tr_3 to give a control waveform swinging (potentially) between ±15 V, although the positive excursion only reaches V_{clamp}. This waveform is routed to control the FET in the sweep circuit, line 1. A_2 and A_1 provide currents via R_6 and R_5 which are fed to a summing amplifier to provide the main and fine tuning controls, line 2. R_4 is adjusted to make the full range of the centre-frequency set control R_1 just cover the required 30 V varactor tuning range of the TV tuner. Lines 1 and 2 are connected as shown in Figure 4.13(b), line 1 operating the clamp transistor Tr_4. Being a JFET, the gate turns on at 0.6 V above V_{clamp}, so line 1 never in fact reaches +15 V. The sweep generator operates as in Figure 4.12(a), with one or two additions. S_{1A} selects a size capacitor appropriate to the sweep duration, two of the capacitors being reused by altering the charging current by a factor of 100, by means of S_{1C}. (Note that for a linear sweep, it is sufficient to ensure that the ratio of R_{19} to R_{21} is the same as the ratio of the two resistors connected to the non-inverting input of A_3; the actual values can be whatever is convenient.) R_{17} is adjusted so that the ramp output from A_3 swings equally positive and negative about earth. S_2 selects the span from full span for the selected band of operation of the TV tuner, via decade steps down to zero span, where the tuner operates unswept at the spot frequency selected with centre frequency controls R_1 and R_2. R_{16} provides a continuously variable control between the settings given by S_2. R_{14} enables the full span (with VAR at max.) to be set to just swing over the 0 to 30 V tuning range of the tuner when centre frequency R_1 is set appropriately. (If centre frequency is set to minimum or maximum,

164 Analog circuits cookbook

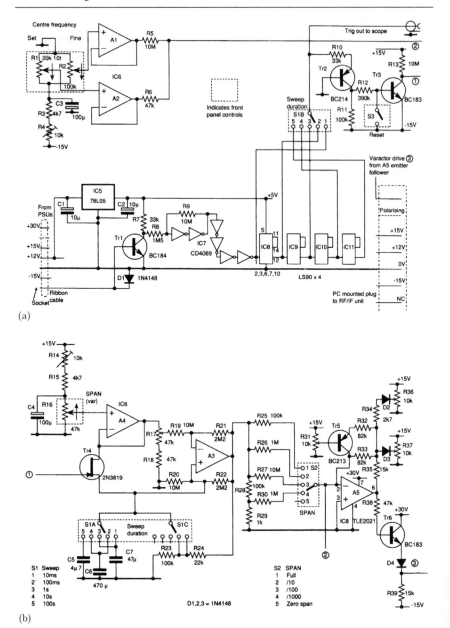

Figure 4.13 (a) Sweep duration generator and centre-frequency setting circuits. (b) Sweep generator, sweep/centre-frequency summer and sweep shaping circuits. NOTE: C_6 and C_7 are 470n and 47n respectively

only the upper or lower half of the span will be displayed, at the left or right side of the oscilloscope trace respectively.)

Inverting amplifier A_5 sums the negative-going sweep waveform and the negative tuning input from R_1 and R_2, to provide a positive-going voltage between 0 and $+30$ V. It also provides waveform shaping, the reason for which is discussed later. The shaped sweep output from A_5 is level shifted by Tr_6 and D_4 before passing to the TV tuner varactor tuning input, since it is important that the sweep should start right from zero volts if the bottom few MHz of Band I are to be covered. All of the front panel controls shown in Figure 4.13 (except the reset control, of which more later) were mounted on a subpanel behind the main panel and connected to the sweep circuit board – mounted on the same subpanel – via ribbon cable, making a self-contained subunit.

RF section

Figure 4.14 shows the RF/IF unit, which is powered via a ribbon cable from the sweep circuit board. The gain of the TV tuner IC_8 can be varied by means of R_{41}, which thus substitutes for the input attenuator of a conventional spectrum analyser. Compared with the

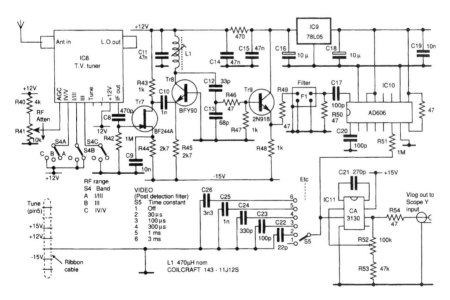

Figure 4.14 *Circuit diagram of the RF/IF unit. This is built around a Toshiba EG522F TV tuner, though almost any other model covering Bands I to V inclusive could be used*

latter, this spectrum monitor has the advantage of a tuned front end, as against a wide open straight-into-the-first-mixer architecture. The front end tuning helps to minimise spurious responses – always a problem with any receiver, including spectrum analysers. The IF output of the tuner, covering approximately 34–40 MHz, is applied via a FET buffer to grounded base amplifier Tr_8. This provides IF gain and some selectivity, its output being buffered by emitter follower Tr_9 and applied to the main IF filter F_1, of which more will be said later. The output of the filter is applied to a true logarithmic IF amplifier of the successive detection variety, the IC used being that featured in Chapter 1, 'Logamps for radar – and much more' (see also Hickman, 1992a).

The required well-decoupled +5 V supply is produced locally by IC_9. The logamp output V_{log} is applied to an output buffer opamp IC_{11} via a simple single-pole switchable video (post-detection) filter, which is useful in reducing 'grass' on the baseline when using a high dispersion (very narrow span) and a suitably low sweep speed. Filter time constants up to 1 s were fitted in the instrument illustrated, but such large values will only be useful with wide dispersions at the slowest sweep speeds. The buffered V_{log} is applied to the Y input of the display used, typically an oscilloscope. R_{52} permits the scaling of the output to be adjusted to give a 10 dB/div. display.

Special considerations

The frequency versus tuning voltage law of the TV tuner is not linear, being simply whatever the *LO* varactor characteristic produces. Just how non-linear is clearly shown in Figure 4.15(a) which shows both the linear tuning ramp and the output V_{log} from the IF strip, showing harmonics of a 10 MHz pulse generator at 50, 60, ..., 110 MHz plus a 115 MHz marker (span range switch S_2 being at full span and span variable control R_{16} fully clockwise). Also visible are the responses to the signals during the retrace, these being telescoped and delayed. The frequency coverage is squashed up in the middle and unduly spread out towards the end – with a yawning gap between 110 and 115 MHz.

The result of some simple linearisation is shown in Figure 4.15(b). As the ramp reaches about 10 V, Tr_5 turns on, adding a second feedback resistor R_{32} in parallel with R_{33}, halving the gain of A_5 and slowing the ramp down so as to decompress the frequency coverage in the region of 70 to 100 MHz, maintaining a 10 MHz/div. display. Just before 100 MHz, D_2 turns on, shunting some of the feedback current via R_{35} away from the input and thus speeding the ramp up again, whilst another more vicious break point due to D_3 at around 110 MHz

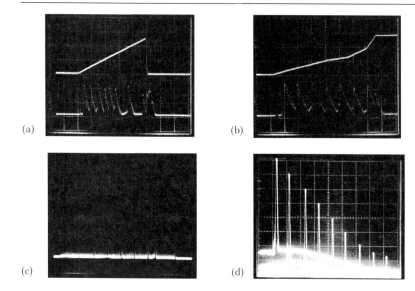

Figure 4.15 *(a) Upper trace, channel 1: the sweep output at cathode of D_4 before the addition of linearising circuitry, 2ms/div. horizontal, 10 V/div. vertical. Lower trace, channel 2: output V_{log} from IF strip showing harmonics of a 10 MHz pulse generator at 50, 60 ..., 110 MHz plus a 115 MHz marker. Sweep time 10 ms. (b) Upper trace: the ramp after shaping to linearise the frequency coverage, 1 ms/div. horizontal, 10 V/div. vertical. Lower trace: as (a). Note that as the ramp now reaches +30V in less than the 10 ms nominal sweep time, the response during the retrace are off-screen to the right. (c) Channel 2 only: as (b) except sweep time 100 ms. Many FM stations now visible in the range 88–104 MHz. (d) 80 MHz CW signal reducing in six steps of 10 dB plus two further steps of 5 dB. Indicating excellent log-conformity over a 65 dB range. SWEEP 100 ms, SPAN 300 kHz/div., VIDEO FILTER 100 μs. (For clarity, the spectrum monitor fine tuning control was used to offset the display of the signal one division to the right at each step in this multiple exposure photo)*

speeds the ramp on its way to 30 V, correctly locating the 115 MHz marker just half a division away from the 110 MHz harmonic. The linearisation has been optimised for operation on Band A (Bands I and II) and holds quite well on B (Band III) with the particular tuner used. Ideally other shaping stages similar to A_5 would be employed for Band III and Band IV/V.

Note that whilst the linearisation shown in Figures 4.14 and 4.15 has produced an approximately constant 10 MHz/div. display on full span, for reduced spans S_2 attenuates the sawtooth *before* it is fed to the shaping stage. Consequently, for reduced spans of the actual span/div. depends upon the setting of the centre-frequency control, although the portion of the full band displayed will be approximately linear, except where it happens to lie across one of the break points.

The filter used in the spectrum monitor illustrated is a 35.4 MHz 6 pole crystal filter designed for 20 kHz channel spacing applications. However, this filter is not ideal, having a basically square passband shape approximating the proverbial brick wall filter. This is not a great inconvenience in practice: it simply means that a slower sweep speed than would suffice with an optimum Gaussian filter must be used. Even for such an optimum filter, the combination of large span and fast sweep speed used in Figure 4.15(a) and (b) would have been quite excessive – it was used as the stretching of the responses makes the effect of linearisation more easily visible. Figure 4.15(c) shows the same Band A (43–118 MHz) display using the nominal 100 ms sweep. FM stations in the range 88 to 104 MHz are clearly visible, no longer being lost in the tails of other responses.

Although the particular crystal filter used is no longer available, a number of alternatives present themselves. A not too dissimilar filter with a centre frequency of 34.368 MHz is available from Webster Electronics (see Ref. 8). Its 20 kHz 3 dB bandwidth (compared with 9.5 kHz for the filter used in the prototype) would permit faster sweep speeds or wider spans to be used, but being only a 4 pole type its ultimate attenuation is rather less, and the one-off price makes it unattractive. A choice of no fewer than five crystal filters in the range 35.0–35.9 MHz is available from Inertial Aerosystems (see Ref. 5), with bandwidths ranging from 8 kHz at –6 dB (type *XF-354S02*) to 125 kHz at –3 dB (type *XF-350S02*, a linear phase type).

A simple alternative is to use synchronously tuned *LC* filters as in the design of Wheeler (1992), though at least twice as many tuned stages should be employed in order to take advantage of the greatly increased on-screen dynamic range offered by the logamp in the design featured here, compared with the linear scale used by Wheeler. The excellent dynamic range of the spectrum monitor is illustrated in the multiple exposure photo, Figure 4.15(d), which shows an 80 MHz CW signal applied to the monitor via a 0–99.9 dB step attenuator. The signal generator output frequency and level were left constant and a minimum of 20 dB attenuation was employed, to buffer the monitor input from the signal generator output. The attenuation was increased by 60 dB in 10 dB steps and then by two further steps of 5 dB, the display of the signal being offset to the right using the centre-frequency controls at each step. Figure 4.15(d) shows the excellent log-conformity of the display over a 65 dB range, the error increasing to 3 dB at –70 dB relative to top-of-screen reference level. It also shows the inadequate 63 dB ultimate attenuation of the crystal filter used, with the much wider LC stage taking over below that level.

An alternative to crystal or LC filters is to use SAW filters, a suitable type being Murata *SAF39.2MB50P*. This is a low impedance 39.2 MHz type designed for TV/VCR sound IF, some additional gain being necessary to allow for its 17 dB typical insertion loss. Two of these filters (available from INTIME Electronics; see Ref. 4) would provide an ultimate attenuation of around 80 dB, enabling full use to be made of the subsequent logamp's dynamic range. The 600 kHz 6 dB bandwidth of each filter would limit the discrimination of fine detail, but allow full span operation at the fastest sweep speed. They could then be backed up by switching in a narrower band filter, e.g. a simple crystal filter.

Using the spectrum monitor

In use, this spectrum monitor is rather like the earliest spectrum analysers; that is to say it is entirely up to the user to ensure that an IF bandwidth (if a choice is available), video filter setting and sweep speeds are used which are suitable for the selected span. Slightly later models had a warning light which came on if the selected IF and/or video filter bandwidth were too narrow, informing the user that wider filter bandwidth(s) should be used, the span reduced or a slower timebase speed employed. Failure to do so means that as the spectrum analyser sweeps past a signal, the latter will not remain within the filter bandwidth long enough for its full amplitude to be registered. This is particularly important in a full-blown spectrum analyser, where the reference level (usually top of screen) is calibrated in absolute terms, e.g. 0 dBm. Later models still, such as the *HP8558B*, had the span and IF bandwidth controls mechanically interlocked, although they could be uncoupled as a convenience for those who knew what they were doing and a snare for those who did not. Both of these controls plus the video filter were also interlocked electrically with the time/div. switch, provided the latter was in the AUTO position. The other positions covered a wide range of different sweep speeds, providing yet further opportunities for the inexperienced to mislead themselves.

Modern spectrum analysers have a microcontroller firmly in command, making the instrument as easy to drive as a modern automatic saloon. By comparison, the spectrum monitor here presented is a veteran car with a crash gearbox, manual advance/retard and rear wheel brakes only – but it will still help you to get around your rf circuitry faster than Shank's pony, as the following examples show.

170 *Analog circuits cookbook*

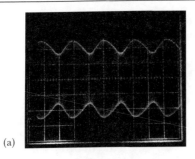

(a)

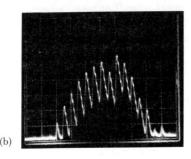

(b)

(c)

Figure 4.16 *(a) Oscilloscope display of the 100 MHz output at maximum level from an inexpensive signal generator, with the fixed level internal 1 kHz AM applied. Oscilloscope set to 100 mV/div. vertical, 500 µs/div. horizontal. (b) Display using the spectrum monitor of the same output but using 50 kHz external modulation depth. SPAN 100 kHz/div. 10 ms SWEEP speed. (c) As (b), but external modulation input reduced by 30 dB, displayed 100 ms SWEEP*

Figure 4.16(a) shows the 100 MHz output from an inexpensive signal generator of Japanese manufacture, with the internal 1 kHz amplitude modulation switched on. The modulation is basically sinusoidal, though some low order distortion is clearly present.

50 kHz external sinusoidal modulation was applied in place of the internal modulation, adjusted for the same modulation depth. Figure 4.16(b) shows the output, this time displayed via the spectrum monitor, at a dispersion of 100 kHz/div. The large number of sidebands present, of slowly diminishing amplitude, are much more than could be explained by the small amount of AM envelope distortion, indicating a great deal of incidental FM on AM, a common occurrence in signal generators when, as here, the amplitude modulation is applied to the RF oscillator stage itself.

In Figure 4.16(c), the amplitude of the applied 50 kHz modulating waveform has been attenuated by 30 dB, so the AM modulation depth is reduced from about 20% in Figure 4.16(a) to 0.63%. This corresponds to AM sidebands of about 50 dB down on

Measurements (rf) 171

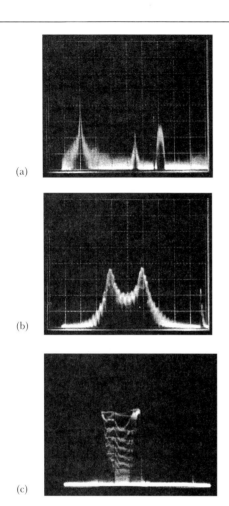

(a)

(b)

(c)

Figure 4.17 (a) A Band IV TV signal, showing (left to right) the vision carrier, colour subcarrier, sound subcarrier and Nicam digital stereo signal. (b) 4.8 kbit/s data FSK modulated onto a VHF carrier; 10 dB/div. vertical, 40 kHz/div. horizontal. (c) High modulation index FM produced by a triangular modulating waveform has a near rectangular envelope with a flat top and steep sides. Individual spectral lines are not visible in this 20 s exposure as there was no clear relation between the modulating frequency and the sweep repetition period. The wavy lines are due to ringing on the tails of the filter response

carrier, whereas those in Figure 4.16(c) are only around 30 dB down. They are therefore clearly almost entirely due to FM, the AM sidebands being responsible for the slight difference in level between the upper and lower FM sidebands. (Whilst AM and first FM sidebands on one side of the carrier add, those on the other subtract.) Note that at the 10 ms sweep used in Figure 4.16(b) the sidebands are not completely resolved. For Figure 4.16(c), the 100 ms sweep was selected, the 50 kHz sidebands being resolved right down to the 60 dB level.

Figure 4.17(a) shows the spectrum monitor operating on Band C – covering bands IV and V. The span is just over 1 MHz/div. and shows a band IV TV signal showing (left to right) the vision carrier, the colour subcarrier, the sound carrier and immediately adjacent to it, the much broader band occupied by the NICAM sound channel.

Figure 4.17(b) shows a 4.8 kbit/s data applied to a VHF FM modulator, producing FSK with a ±40 kHz shift. The signal is spread over a considerable band and clearly a receiver bandwidth in excess of 80 kHz would be necessary to handle the signal. If a carrier is frequency modulated with a sinewave using a very large modulation index (peak deviation much larger than the modulating frequency), a rather similar picture results, except that the dip in the middle is much less pronounced and the sidebands fall away very rapidly at frequencies beyond the peak positive and negative deviation.

The spectrum shape approximates in fact the PSD (power spectral density) of the baseband sinewave. The PSD of a triangular wave is simply rectangular, and Figure 4.17(c) shows triangular modulation applied to the inexpensive signal generator. At the carrier frequency of 100 MHz, the 'AM' modulation is in fact mainly FM and clearly closely approximates a rectangular distribution, the variation being no more than ±1 dB over a bandwidth of 100 kHz.

Such a signal is a useful excitation source for testing a narrowband filter; the filter's characteristic can be displayed when applying its output to a spectrum analyser. This technique is useful when, as with this spectrum monitor, there is no built-in tracking oscillator. A modulating frequency which bears no simple ratio to the repetition rate of the display sweep should be used, otherwise a series of spectral lines, stationary or slowly passing through the display, may result. This is due to a stroboscopic effect similar to the stationary or slowly rolling pattern of a Lissajous figure when the two frequencies are at or near a simple numerical relation.

Further development

A number of refinements should be incorporated in this spectrum monitor to increase its capabilities and usefulness. One simple measure concerns the method of display. As my oscilloscope has a wide range of sweep speeds in 1–2–5 sequence plus a variable control, the output from S_{1B} was simply used as a scope trigger. However, if R_{16} is set permanently at V_{clamp} and a further buffer opamp added between A_3 and A_5 to implement the SPAN(VAR) function, the fixed amplitude output from A_3 (suitably scaled and buffered) can be fed out to the display oscilloscope, set to dc coupled external X input, providing a sweep speed automatically coupled to the sweep speed control S_1. At the slower sweep speeds, e.g. 1 or 10 seconds per sweep, a long persistence scope provides better viewing, whilst for the 100 s sweep a digital storage scope or a simple storage adaptor such as the Thurlby-Thandar *TD201* is very useful – I use the rather more sophisticated Thurlby-Thandar *DSA524* storage adaptor.

However, the slower sweep speeds are only necessary when using a narrow filter with a wide span; for many applications the 10 or 100 ms sweep speed will prove adequate, the 100 ms sweep being acceptable with an oscilloscope using the usual P31 medium–short persistence phosphor. If one of the slowest sweep speeds is in use, it can be very frustrating to realise just after the signal of interest appears on the screen that one needed a different setting of this or that control, since there will be a long wait while the scan completes and then restarts. Pressing the reset button S_3 will reset the tuner sweep voltage to V_{clamp} to give another chance to see the signal, but without resetting either the sweep period selected by S_1 or the oscilloscope trace. If one of the sections of the *CD4069 IC_7* is redeployed to a position between S_{1B} and R_{10}, the sweep will occur during the negative half of the squarewave selected by S_{1B} (see Figure 4.12(b)). A second pole of S_3 can then be used to reset IC_{8-11} to all logic zeros, avoiding a long wait during the unused 50% of the selected squarewave output from S_{1B} before the trace restarts – assuming the display scope is in the external X input mode, rather than using triggered internal timebase.

Working with a single IF bandwidth has its drawbacks; with wide spans a slow sweep speed must be used if the full amplitude of each response is to be measured, whilst with narrow spans the resolution is likely to be insufficient to resolve individual sidebands of a signal. On the other hand, switching filters is a messy business, however it is achieved. Figure 4.18 shows an economical and convenient scheme, using inexpensive stock filters. *LC* or *SAW* filters operating

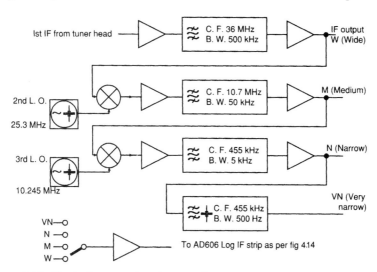

Figure 4.18 *Block diagram showing modified architecture giving a choice of IF bandwidths. It is simpler to provide different signal paths for the different bandwidths rather than select the bandwidth by switching in one or other of several filters all operating at the same IF frequency*

somewhere in the range 35–39 MHz are used for the first IF, providing a wide IF bandwidth permitting full span on each band to be examined without resort to very slow sweep speeds. A conversion to 10.7 MHz enables inexpensive stock 50 kHz filters (e.g. Maplin type number *UF71N*, as used for fast sweeping in scanners) to be used as an intermediate bandwidth, whilst a further conversion to 455 kHz provides a choice of filters with bandwidths of 5 kHz or less from suppliers such as Cirkit, Bonex, etc. As Figure 4.18 indicates, no filter switching is involved: the desired output is simply selected and fed to the log IF strip, which can operate quite happily at each of these frequencies. The net gain of the second and third IFs is fixed at unity, so that switching bandwidths does not alter the height of the displayed response – provided of course that the span and sweep speed are not excessive. Crystals of the appropriate frequencies for the 2nd and 3rd local oscillators can be ordered via an economical 'specials' service (McKnight Crystals; see Ref. 6.)

Another improvement would be better linearisation of the frequency axis, avoiding sharp break points, with the provision of shaping appropriate to each band. The easiest way to achieve this is probably to store *n* values in PROM, *n* being a power of two, and read these out successively to DAC. The *n* values would correspond to equal increments along the frequency axis, each value being what was required to provide the appropriate tuning voltage from the DAC. Chapter 9 of Hickman (1993) described a method (using multiplying DACs) of linearly interpolating between points, giving in effect a shaped varactor drive voltage waveform with *n* break-points per scan. With many break-points available, the change of slope at each will be very small, avoiding the harsh breaks visible in Figure 4.15(b). The two MSBs of the PROM could be used as select lines to call up a different law for each of the three bands.

A very useful feature is a frequency readout, indicating the frequency at the centre or any other point of the display. A true digital readout can be provided by counting the frequency of the *LO* output from the TV tuner, prescaled by a divide-by-100 circuit (see, for example, Hickman, 1992b) to a more convenient frequency. Using the positive half cycle of the 5 Hz squarewave at pin 12 of IC_8 provides a 100 ms gate time which, in conjunction with the divide-by-100 prescaler, gives a 1 kHz resolution. The positive-going edge can be used to jam a count equal to one-hundredth of the IF frequency into a string of reversible counters, set to count down, the appearance of the borrow output switching a flip-flop to set the counters up to count for the rest of the gate period.

The negative-going edge can reset the flip-flop and latch the count; for economy the negative half period could simply enable a seven segment decoder/display driven direct from the counters if you don't mind a flashing display. If span is set to zero, the tuned frequency is indicated exactly. If span is set to one-thousandth or even one-hundredth of full span, the frequency will correspond to the centre of the screen, being of course the average frequency over the duration of the scan. In principle, the same applies up to full span, if the linearisation is good.

A simpler scheme for frequency readout uses an inexpensive DVM. The output of A_2, besides feeding A_5, is also fed to a summing amplifier with presettable gain, which combines it with a presettable offset. This is arranged (for example, on Band A) so that with R_1 at zero, its output is 430 mV and with R_1 at maximum its output is 1.18 V. This is fed to the DVM on the 2.000 V FSD range, providing a readout of 100 kHz/mV. Similar scaling arrangements can be employed for the other bands, the accuracy of the resulting readout depending upon the accuracy of the linearisation employed. This arrangement ignores the effect of centre-frequency fine control R_2, which can if desired be taken into account as follows. The outputs of A_1 and A_2 are combined in a unity gain non-inverting summing amplifier, the output of which is fed via a 47 kΩ resistor to A_5 as now, and also to the scaling-cum-offset amplifier.

However, the simplest frequency calibration scheme of all, unlike the counters and displays, requires no additional kit whatever and, unlike the DVM scheme, is totally independent of the exactness of linearisation. It is simply to calibrate, for each of the three bands, the centre screen (or zero span) frequency against the reading of the digital dial of the ten turn set centre-frequency control potentiometer R_1. Calibration charts have been out of fashion since the days of the *BC212*, but they are as effective as they are cheap, and in the present application they can also be very accurate, since all of the instrument's supplies are stabilised.

My final word concerns not so much an improvement to the existing design as a major reorganisation, but one promising very real advantages. It partly fills in the missing coverage between the top of Band III and the bottom of Band IV, while also adding coverage from 0 Hz up to the bottom of Band I. With the trend to drift-free phase-locked tuning in modern TVs, many tuners now available will probably, like the Toshiba *EG522F*, have an *LO* output available. Figure 4.19(a) shows the *LO* output from the tuner when tuned near the bottom of Band IV/V. The level of the 490 MHz fundamental is −18 dBm and the second and third harmonics are both well over 25 dB down. The *LO* output over the rest of the band is well in excess of −18 dBm. Using broadband amplifiers to boost the tuner's *LO* output to say +7 dBm, it can then be applied as the mixer drive to a

176 *Analog circuits cookbook*

commercial double-balanced mixer, the signal input being applied to the mixer's signal port via a 400 MHz lowpass filter. This tuner is used purely as a local oscillator, the mixer's output being applied to the signal input of a second TV tuner, fixed tuned to 870 MHz (Figure 4.19(b)). The second tuner thus becomes the first IF of an upconverting 0–400 MHz spectrum analyser, its output being applied to a 35 MHz second IF strip as in Figure 4.14. This arrangement provides continuous coverage from 0 Hz almost up to the top end of the 225–400 MHz aviation band in one sweep, so only one set of sweep linearisation is necessary. A most useful feature in a spectrum analyser, not always found even in professional models, is a tracking generator: this provides a constant amplitude cw test signal to which

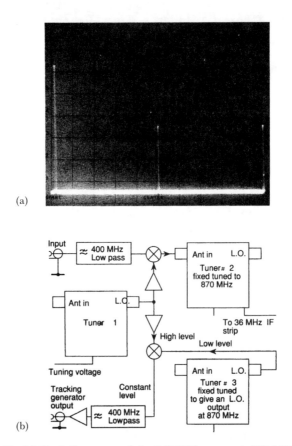

Figure 4.19 *(a) The LO output of the EG522F tuner at 490 MHz, showing also the second and third harmonics. Span 100 MHz/div. vertical 10 dB/div., ref. level (top of screen) 0 dBm. (b) Block diagram of a spectrum monitor based on TV tuners, providing continuous coverage from below 1 MHz up to approx. 400 MHz*

the analyser is always on tune. Figure 4.19(b) also shows how, for the small cost of yet another tuner and mixer, such a facility can be engineered. Used in conjunction with a reflection coefficient bridge, it turns a spectrum analyser into a rudimentary scalar network analyser.

References

1. Hickman, I. (1992a) Logamps for radar – and much more. *Electronics World + Wireless World*, April, 314.
2. Hickman, I. (1992b) A low cost 1.2 GHz pre-scaler. *Practical Wireless*, August, 18–23.
3. Hickman, I. (1993) *Analog Electronics*. Butterworth-Heinemann, Oxford.
4. INTIME Electronics Ltd. Tel. 01787 478470.
5. KVG-GMBH, UK agent: Inertial Aerosystems Ltd. Tel. 01252 782442.
6. McKnight Crystals Ltd. Tel. 01703 848961.
7. SENDZ Components, 63 Bishopsteignton, Shoeburyness, Essex SS3 8AF. Tel. 01702 332992.
8. Tele Quarz type TQF 34-01, Webster Electronics. Tel. 0146 05 7166.
9. Wheeler, N. (1992) Spectrum analysis on the cheap. *Electronics World + Wireless World*, March, 205.

A wideband isolator

Circulators and isolators are linear non-reciprocal signal handling components, with a number of uses at rf. They have something in common with directional couplers – indeed they are a type of directional coupler, but with intriguing properties. Circulators and isolators are common components at microwave, but large and expensive at UHF and just not available at lower frequencies. At least, that was the case until recently.

Wideband isolator

Circulators and isolators

Circulators and isolators are examples of directional couplers, and are common enough components at microwave frequencies. They are three port devices, the ports being either coaxial or waveguide connectors, according to the frequency and particular design. The clever part is the way signals are routed from one port to the next, always in the same direction. The operation of a circulator (or isolator) depends upon the interaction, within a lump of ferrite, of the rf field due to the signal, and a steady dc field provided by a permanent magnet – something to do with the precession of electron orbits, or so I gather from those who know more about microwaves. They can be used for a variety of purposes, one of which is the subject of this article.

Figure 4.20(a) shows (diagrammatically) a three port circulator, the arrow indicating the direction of circulation. This means that a signal applied at port A is all delivered to port B, with little (ideally none, if the device's 'directivity' is perfect) coming out of port C. What happens next depends upon what is connected to port B. If this port is terminated with an ideal resistive load equal to the device's characteristic impedance (usually 50 Ω in the case of a circulator with coaxial connectors), then all of the signal is accepted by the termination and none is returned to port B – the 'return loss' in dB is infinity. But if the termination on port B differs from (50 + j0) Ω, then there is a finite return loss. The reflected (returned) signal goes back into port B and circulates around in the direction of the arrow, coming out at port C. Thus the magnitude of the signal appearing at port C, relative to the magnitude of the input applied to port A is a measure of the degree of mismatch at port B. Thus with the aid of a source and detector, a circulator can be used to measure the return loss – and hence the VSWR – of any given DUT (device under test), as in Figure 4.20(b). This rather assumes that the detector presents a good match to port C. Otherwise it will reflect some of the signal it receives, back into port C – whence it will resurface round the houses at port A.

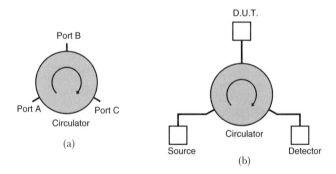

Figure 4.20 (a) A three port circulator. (b) An arrangement using a circulator to measure the return loss of a device under test

Given a total mismatch (a short or open at port B), then all of the power input at port A will come out at port C (but strictly via the clockwise route) – bar the usual small insertion loss to be expected of any practical device. Because it is a totally symmetrical device, the circulator in Figure 4.20(b) could be rotated by 120 or 240° and still work exactly the same. It matters not which port the source is connected to, provided the DUT and detector are connected to the following two in clockwise order. An isolator is a related, if less totally symmetrical, device. Here, any signal in Figure 4.20(b) reflected back into port C by the detector is simply absorbed, and not passed around back to port A. Thus an isolator would actually be a more appropriate device for the VSWR measuring set-up of Figure 4.20(b), though for some other applications circulators might be preferred.

Microwave circulators with high directivity are narrow band devices. Bandwidths of up to an octave are possible, but only at the expense of much reduced directivity. Circulators and isolators are such useful devices, that it would be great if economical models with good directivity were available at UHF, VHF and even lower frequencies. And even better if one really broadband model were available covering all these frequencies at once.

The answer to a long felt want

Though not as well known as it deserves, such an arrangement is in fact possible. It filled me with excitement when I first came across it, in the American controlled circulation magazine *RF Design*, Ref. 1. This circuit uses three CLC406 current feedback opamps (from Comlinear, now part of National Semiconductor), and operates up to well over 100 MHz, the upper limit being set by the frequency at which the opamps begin to flag unduly.

What the article describes is nothing less than an active circuit switchable for use as either a circulator or an isolator, as required. It has three 50 Ω BNC ports, and operates from – say – 200 MHz, right down to dc. The circuit is shown in Figure 4.21.

Whilst at the leading edge of technology when introduced, and still a good opamp today, the *CLC406* has nonetheless been overtaken, performance-wise, by newer devices. In particular, the *AD8009* from Analog Devices caught my interest, with its unity-gain bandwidth (small signal, non-inverting) of 1 GHz. Of course, if you demand more gain or apply large signals, the performance is a little less – 700 MHz at a small signal gain (0.2 V pp) of +2, or 440 MHz, 320 MHz at large signal gains (2 V pp) of +2, +10. Still, it seemed a good contender for use in an updated version of the circuit described above. But before

180 Analog circuits cookbook

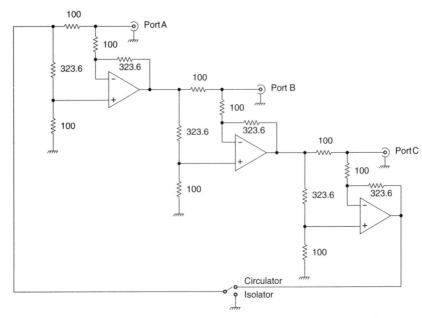

Figure 4.21 *The circuit of the active circulator/isolator described in Ref. 1*

going on to describe it, it might be as well to analyse the circuit to show just how it works.

How it works

A feature of this circuit is that it works down to dc. So its operation can be described simply with reference to the partial circuit shown in Figure 4.22. Here, the voltages may be taken as dc, or as ac in-phase (or antiphase where negative).

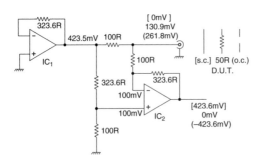

Figure 4.22 *Partial circuit, explaining circuit operation*

Instead of assuming an input voltage and trying to derive the output voltage, or vice versa, a useful trick in circuit analysis is to assume a convenient voltage at some internal node, and work forwards and backwards from there. The results then drop out fairly simply, even by mental arithmetic in some cases. So let's assume the voltage at the + input (non-inverting) of IC_2 is 100 mV. Then the voltage at the output of IC_1 must be 423.5 mV. Also, due to the negative feedback, IC_2's output will do whatever is necessary to ensure that its – input (inverting) is also at 100 mV.

Figure 4.22 shows what the output of IC_2 will be, for the cases of a short circuit [s.c.], or 50 Ω, or an open circuit (o.c.) at the port. The s.c. case is obvious: the resistor at IC_2's inverting input and its feedback resistor form an identical chain to that at the + input. Thus the output of IC_2 is at +423.6 mV, like IC_1, the overall gain is +1, but note that the opamp is working at a gain in excess of +3. In the o.c. case, the net volt drop across the two 100 Ω resistors in series is 323.6 mV, so the output of IC_2 must be at (323.6/200) 323.6 mV negative with respect to the – input, which works out, thanks to the careful choice of resistor values, at –423.6 mV.

With a 50 Ω termination at the port, a line or two of algebra on the back of an envelope may be needed. Let the voltage at the port be v. Now equate the current flowing from IC_1 output to the port, to the sum of the currents flowing from there to ground via 50 Ω and to the inverting input of IC_2 via 100 Ω. v drops out immediately, defining the current flowing through the input and feedback resistors of IC_2, and hence the voltage at IC_2's output. It turns out (again thanks to the ingenious design of the resistive network between each of the opamps) that the voltage at the output of IC_2 is zero and the corresponding voltage at the DUT port is 130.9 mV. Since this is precisely the voltage at a port which produces 423.5 mV at the output of the following opamp, clearly it is the voltage that must be applied to the source input port A (not shown in Figure 4.22) which drives IC_1. Hence the gain from port A to B (or B to C, or C to A) is unity, provided that both the two ports 'see' 50 Ω. And if the second port sees an infinite VSWR load, the gain from the first to the third port is unity. Effectively, all the power returned from the second port circulates round to the third. At least, this is the case with a circulator. As Figure 4.21 shows, in the case of an isolator, any incident power reflected back into port C is simply absorbed, and does not continue around back to port A.

An updated version

With wideband current feedback opamps type *AD8009* from Analog Devices available, I was keen to see what sort of performance could nowadays be achieved. Clearly, they could simply be substituted for the *CLC406* in the circuit of Figure 4.21. However, after careful consideration, it seemed that all the applications I had in mind could be met with an isolator. Now if one is willing to forego the ability to

182 Analog circuits cookbook

switch the circuit to operate, when required, as a circulator, then not only are substantial economies in circuit design possible, but furthermore, one or two dodges to improve performance at the top end of the frequency range can be incorporated.

So at the end of the day, my circuit finished up as in Figure 4.23. It can be seen straight away, that as an isolator only, the circuit needs but two opamps. Also not needed are a switch and a number of resistors, while the port C is simply driven by an L pad.

But before describing the operation of the rf portion of Figure 4.23, a word about the power supply arrangements is called for. Circuits under development sometimes fail for no apparent reason. This is often put down to 'prototype fatigue', meaning some form of unidentified electrical abuse. I have suffered the ravages of this phenomenon as often as most.

The construction of the isolator, using opamps in the small outline SO8 package and chip resistors and 0805 10n capacitors, was not a simple task, involving both dexterity and some eye strain! The circuit was built using 'fresh air' construction on a scrap of copperclad FRG used as a groundplane. The thought of having to dive back into the bird's nest to replace an opamp or two was horrific, so some protection for the supplies was built in. The series diodes guard against possible connection of the power supplies in reverse polarity, whilst the zener diodes prevent excessive voltage being applied. The types quoted will not provide indefinite protection from 15 V supplies with a 1 A current limit, but guard against an insidious and often unrealised fault. Some (ageing) lab bench power supplies output, at switch-on, a brief spike of maximum voltage equal to the internal raw supply voltage. And many power supplies, after a number of years' use, develop a noisy track on the output voltage setting pot. This likewise, depending on the particular design, can result in a brief

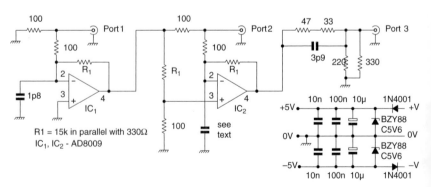

Figure 4.23 *Circuit diagram of a wideband isolator, usable from 0 Hz to 500 MHz*

spike of maximum output voltage whenever the pot is adjusted. For the sake of a few extra components, it is better to be safe than sorry.

Construction

The two opamps were mounted in between the three BNC sockets, which were placed as close together as thought would be possible. In the event, it turned out that they could have been a little closer still, but no matter. In somewhat cavalier style, the ICs were mounted above the groundplane, standing on leads 1, 5 and 8 (also lead 3 in the case of IC_1). These leads had been carefully bent down from the usual horizontal position on a surface mount device, the remaining leads having been bent upwards. A 10n 0805 chip capacitor was then soldered between the ground plane and each supply lead, leaning in towards the device at an angle of about 60° from the vertical. The leaded 100n capacitors (also four in total, these items of Figure 4.23 being duplicated) were then also fitted to each side of the opamp to leave space for the chip resistors. The chip resistors were then fitted, the feedback resistors around IC_1 and IC_2 being mounted on top of the devices, directly between the bent-up pins 2 and 6. As the body length of the 100 Ω input resistor to IC_1 was not sufficient to reach the shortened spill of the BNC centre contact at port A, the gap was bridged by a few millimetres of 3 mm wide 1 thou copper tape, the same trick being used elsewhere, where necessary. (If you don't have any copper tape to hand, a little can always be stripped from an odd scrap of copperclad. The application of heat from a soldering iron bit will enable the copper to be peeled from the board – this is possible with GRP and even easier with SRBP.)

Testing

The finished prototype was fired up and tested, using the equipment briefly described later. Performance up to several hundred MHz was very encouraging, but it was obviously sensible to try and wring the last ounce of performance from the circuit. Figure 4.24(a), reproduced from the *AD8009* data sheet, shows how a useful increase in bandwidth can be achieved by the addition of different small amounts of capacitance to ground from the opamp's inverting input, at the expense of some peaking at the top end of the frequency range. (Figure 4.24(b) shows the effect of those same values of capacitance on the pulse response.) In Figure 4.23, the opamps are used at a gain in excess of +10 dB, so the same degree of bandwidth extension cannot be expected for sensible values of capacitance at the opamp's inverting input.

184 *Analog circuits cookbook*

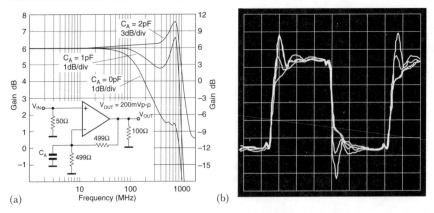

Figure 4.24 *(a) Bandwidth extension for the AD8009 achieved (for a gain of +2) by adding capacitance from the inverting input to ground. (b) The effect of these three values of capacitance on the pulse response*

After some experiment, in the case of IC_1 a value of 1.8p was selected. In the case of IC_2, the value of capacitance was adjusted for best device directivity. This involved terminating port B with a 50 Ω termination and tweaking the capacitance to give the greatest attenuation of the residual signal at port C in the 300–500 MHz region. As the required value was around 1p, lower than the minimum capacitance of the smallest trimmers I had in stock, it was realised as two short lengths of 30 SWG enamelled copper wire twisted together. The length was trimmed back for optimum directivity as described above, leaving just over 1 cm of twisted wire. The transmission path from port A to B and that from port B to C both showed a smooth roll-off above 500 MHz, with no sign of peaking.

Test gear

With such a wideband device, any sensible evaluation of its performance required some form of sweep equipment. Fortunately, the necessary gear, if a little untidy and homemade, was to hand.

For general rf measurements, I have a Hewlett Packard 0.1–1500 MHz spectrum analyser type *8558B*, which is a plug-in unit fitted in a 182T large screen display mainframe. The mainframe and plug-in was purchased as a complete instrument, tested and guaranteed, from one of the dealers in this type of second-hand equipment who advertises regularly in *Electronics World*. Being an older type of instrument, long out of production, it is available at a very modest price (for the performance it offers). Unfortunately, this

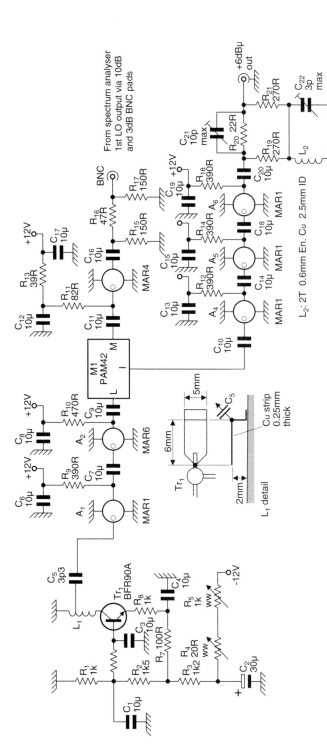

Figure 4.25 Circuit diagram of an appliqué box for an HP 8558B spectrum analyser, providing a 0–1500 MHz tracking generator output. (Reproduced with permission from Electronic Product Design, July 1994, p.17) (Note: for 10μ read 10n)

instrument does not include a built-in tracking generator – those only came in with the introduction of a later generation of spectrum analyser. But it does make a sample of the 2.05–3.55 GHz first local oscillator available at the front panel. Some time ago I published (Ref. 2) a circuit for an add-on for such an instrument. It accepts an attenuated version of the spectrum analyser's first local oscillator output and mixes it with an internally generated CW centred on 2.05 GHz. The output, as the spectrum analyser's first local oscillator sweeps from 2.05 to 3.55 GHz, is a tracking output covering the analyser's 0–1500 MHz input range. The circuit is reproduced here as Figure 4.25.

Testing the isolator's performance

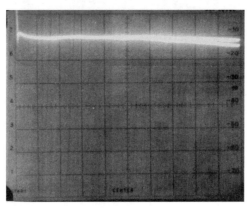

Figure 4.26 *Upper trace, output of the tracking generator, attenuated by 20 dB. Lower trace, as upper trace, but with the signal routed via port B to port C of the isolator. Reference level –2.5 dB, 10 dB/div., span 0–500 MHz, IF bandwidth 3 MHz, video filter medium*

After using the equipment described above to optimise the isolator's performance, some photographs of the screen display were taken for the record. Figure 4.26 shows (upper trace) the output of the tracking generator, connected via two coaxial cables and two 10 dB pads (joined by a BNC back-to-back female adaptor) connected to the input of the spectrum analyser. The sweep covers 0–500 MHz and the vertical deflection factor is 10 dB/div. The back-to-back BNC connector was then replaced by the isolator, input to port B, output from port C. A second exposure on the same shot captured the frequency response of the isolator, Figure 4.26, lower trace. It can be seen that the insertion loss of the isolator is negligible up to 300 MHz, and only about 3 dB at 500 MHz. The response from port A to port B is just a little worse, as this path could not use the frequency compensation provided by the 3.9p capacitor in the output pad at port C.

Figure 4.27 shows the reverse isolation from port B (as input) to port A (lower trace; with the input, upper trace, for comparison). This can be seen to be mostly 45 dB or greater, and better than 40 dB right up to 500 MHz. Given an ideal opamp with infinite gain even at

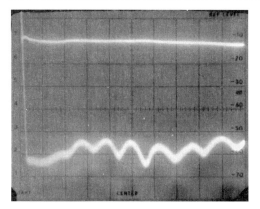

Figure 4.27 Upper trace as Figure 4.26, for reference. Lower trace, output from port A of the isolator with the input applied to port B. Spectrum analyser settings as for Figure 4.26 except video filter at max.

500 MHz, the negative feedback would ensure an effectively zero output impedance. IC_1 would then be able to swallow any current injected into its output from port B with none passing via R_1 to port A.

At lower frequencies this is exactly what happens, the lower trace reflecting in part the limitations of the instrumentation. The fixed 2.05 GHz oscillator Tr_1 in Figure 4.25 is of course running at the analyser's first IF frequency. So any leakage from Tr_1 back into the analyser's first LO output (and thence into the first IF) is by definition always on tune. Indeed, the purpose of R_4, R_5 is precisely to permit tuning of the fixed oscillator (which is not in any way frequency stabilised) to the analyser's first intermediate frequency. The purpose of the external 13 dB pad between the analyser's first LO output and the appliqué box, and the latter's internal pad R_{15}–R_{17} is to minimise this back leakage. Despite these precautions, even with the input to the spectrum analyser closed in a 50 Ω termination, the residual trace due to leakage is only a few dB below that shown in Figure 4.27.

Testing the isolator's directivity

My main use for the isolator is as a means of testing the VSWR of various items of rf kit, such as antennas, attenuators, the input and output impedances of amplifiers, etc. To determine just how useful it was in this role, the output at port C was recorded, relative to the input at port A, for various degrees of mismatch at port B, see Figure 4.28. The top trace is the output level with an open circuit at port B. Comparing it with the upper trace in Figure 4.26, it is about 7 dB down at 500 MHz, this being the sum of the insertion loss from port A to port B, plus the insertion loss from port B to port C – already noted in Figure 4.26 as around 3 dB.

The three lower traces in Figure 4.28 are with a 75 Ω termination at port B (providing a 14 dB return loss), a 50 Ω 10 dB pad open at the far end (providing a 20 dB return loss), and three 50 Ω 10 dB pads

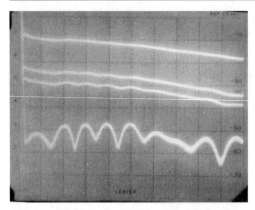

Figure 4.28 *Traces showing the signal at port C for various degrees of intentional mismatch at port B: with (top to bottom) return loss of 0, 14, 20 and 74 dB. Signal applied to port A as in Figure 4.26, upper trace. Spectrum analyser settings as for Figure 4.27*

terminated in 75 Ω. The latter works out as a theoretical 74 dB return loss, or close to 50 Ω, and the resolution of the system as measured is apparently limited to around 40 dB. A return loss of 40 dB corresponds to a reflection coefficient ρ of 1%.

Now $\rho = (Z_t - Z_o)/(Z_t + Z_o)$ where Z_t is the actual value of the termination and Z_o is the characteristic impedance, viz. 50 Ω. So ρ = 1% corresponds to a Z_t of 51 Ω. The dc resistance looking into the string of three 10 dB pads plus the 75 Ω termination was measured at dc as 50.6 Ω. Clearly, then, assuming this is still the case at 500 MHz, much of the residual signal in the bottom trace in Figure 4.28 can be assumed to be due to the error in the characteristic impedance of the pads, which were normal commercial quality, not measurement laboratory standard. For the rest, it is down to the limited directivity of the isolator. To maximise this, the chip resistors were all selected to be well within 1%, from the supply of 5% chips to hand. I had originally hoped to be able to select 326.3 Ω resistors from the 313.5 – 346.5 spread of 330 Ω 5% resistors. But most were in fact within 1%, hence the need for a parallel 15K to secure the right value.

But the interesting, and indeed vital, point is that the directivity of the system does not depend upon the flatness of the frequency response. The fact that the three upper curves in Figure 4.28 are so nearly identical and parallel indicates that the isolator is useful for VSWR measurements right up to 500 MHz, and perhaps a bit beyond. This is because the directivity depends upon two things. Firstly, that the balance of the bridge of resistors at the input of IC_2 in Figure 4.25 remains constant with frequency. Secondly, that the common mode rejection of the opamp remains high right up to 500 MHz, and in view of the excellent results obtained, this certainly seems to be the case.

Using the isolator

The spectrum analyser plus its homebrew tracking generator was very useful for demonstrating the isolator's performance over the whole band up to 500 MHz in one sweep. But the arrangement has its limitations. Apart from the back leakage from the 2.05 GHz oscillator, already mentioned, there are two other limitations. Firstly, as the 0–500 MHz sweep proceeds, the frequency of the 2.05 GHz oscillator tends to be affected slightly, so that it is necessary to use a wider than usual IF bandwidth in the analyser. Secondly, to maintain a sensibly flat output level, the output is taken from an overdriven string of amplifiers, with resulting high harmonic content. This is normally of no consequence, since the analyser is selective and is by definition tuned only to the fundamental. But problems can arise with spurious responses due to the presence of the harmonics.

Where a more modest frequency range, up to 200 MHz, suffices, the sweeper described in Ref. 3 can be used, in conjunction with a broadband detector (perhaps preceded by a broadband amplifier) connected to port C. A successive detection logarithmic amplifier makes a very convenient detector, and types covering frequencies up to 500 MHz are mentioned in Ref. 4.

For many applications, a swept measurement is not essential, e.g. when adjusting a transmitting antenna for best VSWR at a certain frequency. In this case, any convenient signal generator can be used. At the higher frequencies, however, it is best to keep the input to port A to not more than 0 dBm. A receiver can be pressed into service as the detector at port C. Many receivers, e.g. scanners, include an RSSI facility, and in many cases, these make surprisingly accurate log level meters. Measuring the level at port C relative to that at port A will give the return loss, and hence the VSWR, of the DUT connected to port B. Tuning/adjusting it for maximum return loss will provide a DUT with an optimum VSWR. Return loss measurements can be cross-checked at any time by substituting an attenuator(s) and/or 75 Ω termination for the DUT, as described earlier.

Finally, an interesting point about this active circuit. No problem was experienced at any stage with instability. But what about the circulator version of Figure 4.21. Here, any reflected power at port C circulates back around to port A. What happens if all three ports are left open circuit? Given that tolerance variations on the resistors could result in a low frequency gain marginally in excess of unity in each stage, could the circuit 'sing around' and lock up with the opamp outputs stuck at the rail?

In fact the answer is no, because as Figure 4.22 shows, when a port is open circuit, the output of the following opamp is of the opposite polarity. (Thus the voltage passed on to the next stage is of the opposite polarity to the reflected voltage at the stage's input.) Three inverters in a ring are dc stable, and at frequencies where each contributes 60° phase shift or more, the loop gain is already well below 0 dB. Of course, if all three ports are shorted, each stage passes on a (possibly marginally greater) voltage of the same polarity, and lock-up is a possibility. But I can't think of any circumstances where one might want to try and use a circulator with all three ports short circuited!

References

1. Wenzel, Charles. Low frequency circulator/isolator uses no ferrite or magnet, *RF Design*. (The winning entry in the 1991 RF Design Awards Contest.)
2. March, I. (1994) Simple tracking generator for spectrum analyser. *Electronic Product Design,* July, p. 17.
3. Hickman, I. (1995) Sweeping to VHF. *Electronics World,* October, pp. 823–830.
4. Hickman, I. (1993) Log amps for radar – and much more. *Electronics World + Wireless World*, April, pp. 314–317.

5 Opto

> **Linear optical imager**
>
> This item describes an economical optical line imager with 64 point resolution. Applications abound: with some arrangement for vertical scanning, it would even make a rudimentary TV camera, with twice the resolution of Baird's pre-second world war TV system!

Sensing the position

A common requirement in industry, especially with the advance of automation, is position sensing, allied to position control actuators of various kinds. For the simpler jobs, discrete photodetectors, vane switches and the like suffice, but for critical applications, a progressive rather than an on/off indication of position is required. The CCD imaging devices can be used for position sensing, but require several different supplies and auxiliary ICs; additionally the very small pixel pitch (typically 10–15 µm) requires the use of good optics in most applications. The Texas Instruments *TSL214* is a 64 pixel addressed line array with a sensor pitch of 125 µm, giving an active sensor length of 8 mm and permitting the use of cheaper optics. In the *TSL214* (Figure 5.1), the pixel charges are individually switched out sequentially under control of a 64 stage shift register which produces non-overlapping clocks to control this process, unlike a CCD array where all the pixel charges of an integration period are clocked out together down transport registers. The *TSL214* is mounted in an economical 14 pin DIL package with a transparent cover, and the low active pin count makes the production of 128 and 192 pixel devices (*TSL215, TSL216*) a relatively simple process.

192 Analog circuits cookbook

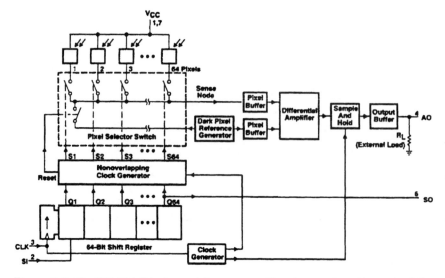

Figure 5.1 *The TSL214 64 element line sensor. The charge stored in each of the 64 pixels during each integration period is read out sequentially via a 64 way mux controlled by the shift register. The charge accumulated in an integration period is proportional to the intensity of the light and the length of the integration period*

Operation of the device is controlled by a clock input (which may be between 10 and 500 kHz) and an SI (serial input) signal which determines the integration time (see Figure 5.2). The integration period includes the 64 clock read-out period, each pixel recommencing integrating immediately after being read out. Consequently, the duration of the minimum integration period is 65 clock periods, though a longer interval between SI pulses (or a lower clock rate) may be used if operation at lower light levels is required.

To gain an insight into their operation, I made up the circuit shown in Figure 5.3. When using the *TSL214*, beware: the end of the package with a semicircular notch is *not* the pin 1 end. The other end has two such notches, and a spot of silver paint over pin 1 which I should have noticed. Having reinserted the device into the circuit the right way round, I found that the output remained stuck at about +3.8 V during the whole of each 64 clock output period, regardless of whether the sensor was covered or not. The obvious conclusion was that the device had been damaged by being inserted back to front. However, there are no internal device connections on the pin 8–14 side except pin 12 (ground), and the corresponding pin on the other side (pin 5) is also ground, rendering the device goof-proof. The problem proved to be the low clock rate of 20 kHz, resulting in a

Opto 193

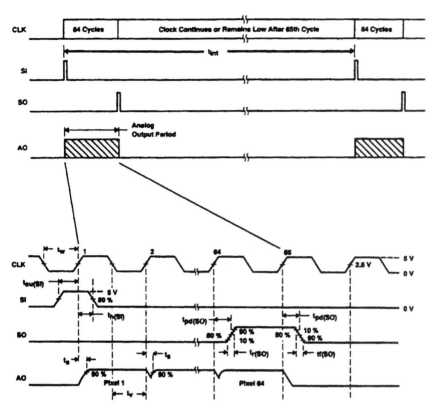

Figure 5.2 *The pixel outputs appear sequentially on the A_o pin following the rising edges of the next 64 clock pulses that follow the assertion of the SI pulse, assuming its set-up and hold times are met. Following the sample time t_s, the pixel analog data is valid for at least the period t_v*

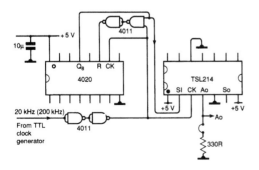

Figure 5.3 *Circuit used for initial investigation of TSL214 operation. An SI (serial input) pulse is produced for every 128th clock pulse. The 64 clock pulses following SI read out the analog light intensity-related signals at pin A_o*

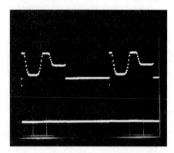

Figure 5.4 *Device operating in the circuit of Figure 5.3. Lower trace: SI pulses; upper trace: analog output at A_o with part of the line array covered*

sensitivity so great that the device could still 'see' the lights over the lab bench through my thumb. Switching the lights off, increasing the clock rate or using a strip of metal to cover the device were all equally effective.

Figure 5.4 illustrates the analog nature of the output. A narrow strip of metal was laid across the device just left-of-centre whilst the right-hand end was covered with a piece of deep green gel the type used with theatre spotlights. The lower trace shows SI pulses and these are immediately followed (upper trace) by 64 analog output samples. Where not covered, the samples are at the maximum output level of just under 4 V; where covered by metal, at the dark level of around 0.2 V and where covered by gel, at an intermediate level. With no optics, the light reaching the device was not collimated but diffuse: consequently light leakage under the edges of the metal strip is clearly apparent in the photograph.

The last stage of the shift register produces an output pulse S_o which can be used to initiate readout from another similar device. To enable device outputs A_o to be bussed up, the A_o output becomes high impedance (tristate) when not outputting samples: clearly this is most useful if the second sensor element is in the same package so that the active area becomes a continuous line, hence the *TSL215* and *216*. This tristate aspect has another use, however. Besides using the device with a microcontroller it can also be used in edge detection and similar applications in a purely analog system. This is illustrated by the circuit of Figure 5.5, where a negative output voltage is produced, proportional

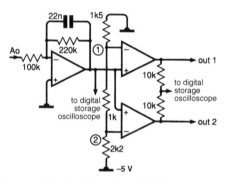

Figure 5.5 *An inverting leaky integrator produces a negative output voltage proportional to the number of pixels which are uncovered. Comparators indicate whether about half the device is illuminated, or more, or less*

Opto 195

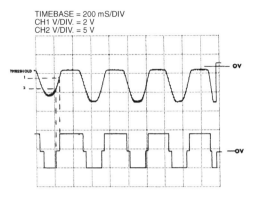

Figure 5.6 *Operating the device in the analog system of Figure 5.5 – 330R load resistor shown in Figure 5.3 removed. Output as a piece of thick card waved back and forth across the sensor (upper trace); output of comparators (lower trace)*

(under conditions of contant incident illumination) to the number cells illuminated.

Figure 5.6 shows this voltage varying as a card is waved back and forth, covering and uncovering the sensor. The voltage could be used to drive a meter and if the meter movement carried a vane which moved across in front of the sensor so as to cover more of it as the output voltage increased, a rather complicated light-meter would result: but with a possibly useful pseudo-logarithmic sensitivity characteristic giving the greatest resolution at the lowest light levels.

Alternatively, the output voltage could be processed as shown by two comparators to indicate that substantially more (or less) than half the array is illuminated. (To show the operation of both comparators on a single trace, their outputs have been combined via 10 kΩ resistors: this is a useful technique for showing two or more simple signals of predetermined format on a single trace.) The comparators could provide steer-left and steer-right commands on a factory robot following the edge of a white line painted on the shop floor. This would provide a bang-bang servo type of control, so it might be better to use the analog voltage directly for steering control, as small deviations would cause only small corrections, resulting in smoother operation.

The *TSL214* features high sensitivity combined with a broad spectral response and low dark current which is almost totally independent of the integration time used (Figure 5.7). It is thus eminently suitable for use in a host of applications, for example a rotary encoder with 1° resolution (Figure 5.8(a)). The output can simply be routed to a microcontroller to provide rotary position information to the host system, or to a display (Figure 5.8(b)).

To assist potential users with initial evaluation of the device, the *PC404* Evaluation Kit is available (Figure 5.9).

This consists of a *TSL214*, a circuit board with drive and output circuitry, and a detachable ×10 magnification lens in a housing. The

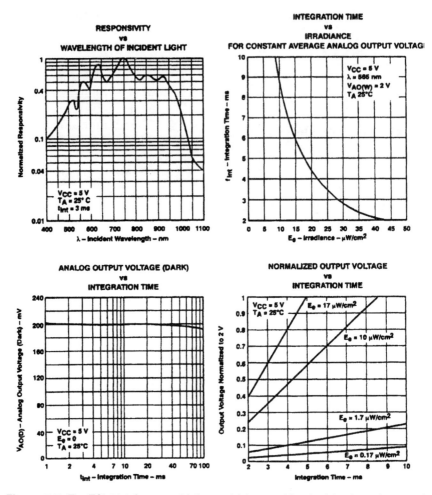

Figure 5.7 *The* TSL214 *features high sensitivity combined with a broad spectral response and low dark current which is almost totally independent of the integration time used*

circuitry of the *PC404* comprises an oscillator, a counter/divider, a one-shot pulse generator and a comparator. The oscillator is built around a *555* timer and generates a 500 kHz output data clock pulse. The clock output of the oscillator is routed to a *74HC4040* divider. This has a set of jumper terminals to four of the outputs, and 1, 2, 4 or 8 ms integration time may be selected. The selected output is connected to the *74HC123* one-shot pulse generator, which provides the *TSL214* with an SI pulse.

Opto 197

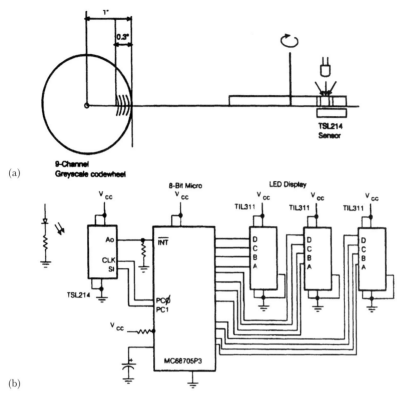

Figure 5.8 *Mated with a 9 channel grey-scale codewheel, the* TSL214 *can provide rotary position readout to better than 1° resolution (a), or display shaft position in degrees (b)*

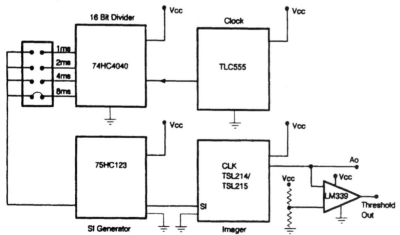

Figure 5.9 *Block diagram of the* PC404 *Evaluation Kit for the* TSL214 *64 pixel integrated array*

Acknowledgements

Several of the illustrations in this article are reproduced by courtesy of Texas Instruments.

> ### Linear optoisolator
>
> Optoisolators are widely used for transmitting simple on/off signals across a galvanic isolation barrier. But the basic optoisolator has a non-linear input–output characteristic and so is not suitable for handling, for example, the feedback signal in a direct-off-mains switching power supply. This article describes a low-drift, high-linearity optoisolator providing the solution.

Bringing the optoisolator into line

A common design requirement is to carry signals across a voltage barrier, e.g. in industrial, instrumentation, medical and communication systems, so that the signal on the output side is entirely isolated and floating relative to the signal on the input side of the barrier. Where the only signal components of interest are ac, simple capacitive isolation may suffice, but often dc coupling is of the essence, as in the control loop of a direct-off-mains switching power supply.

Various schemes have been employed for this purpose, including the use of isolation amplifiers which are available as standard products in IC form from a number of manufacturers, including Analog Devices and Burr Brown. Another method involves the use of a V-to-F (voltage-to-frequency converter) to carry the signal across via a high voltage working capacitor, or via an LED-photodiode link, followed by an F-to-V, but this introduces a delay due to the V-to-F and F-to-V settling times which can introduce an embarrassing phase shift into a control loop. One could of course dispense with the F-to-V and the V-to-F, applying the input signal via a voltage-to-current converter to the LED and taking the output voltage from the coupled photodiode or transistor. The problems here are drift and poor linearity.

A low-drift high-linearity isolator is, however, available in the form of the Siemens *IL300* linear optocoupler. In addition to an LED and a highly insulated output photodiode, the coupler contains a second photodiode which is also illuminated by the LED and can thus be used in a feedback loop to control the LED current. The ratio K_1 of the feedback (servo) photodiode current to the LED current is specified at an LED forward current I_f of 10 mA, as is K_2, the ratio of the output photodiode current to the LED current (Figure 5.10). The two photodiodes are PIN diodes whose photocurrent is linearly related to the incident luminous flux. Consequently, due to the high loop gain of the NFB loop enclosing the LED and the input photodiode, IP_1 in Figure 5.10 will be linearly related to V_{in}, even though the light output of the LED is not linearly related to its forward current.

The constant of linearity is slightly temperature dependent, but this affects the output photodiode equally, so K_3 (the ratio of K_1 and K_2) is virtually temperature independent. Thus $V_o/V_{in} = (K_2R_2)/(K_1R_1) = K_3(R_2/R_1)$. There are production spreads on both K_1 and K_2, and hence also on K_3, so the devices are binned into two selections for K_1 and ten for K_3 (see 'Bin sorting and categories' later in this article) and coded accordingly.

Any semiconductor photodiode can be used in either of two modes, photovoltaic or photoconductive. In the case of the *IL300*, the photoconductive mode provides the higher signal transfer bandwidth and the device's performance is consequently specified in this mode. However, the photovoltaic mode provides lower offset drift and greater linearity (better than 12 bit) so I obtained two sample devices (coded *WI*) for evaluation. The circuit of Figure 5.11 was used to test one of the devices, the scheme being to subtract the output from a sample of the input, leaving only the distortion produced in the device under test. This 'take away the number you first thought of' technique is powerful and useful – within limits. In principle, any test signal will do, but a sinewave is the most useful as it provides information as to the order of the distortion mechanism, if any, in the device under test. A 5 V pp sinewave input at 50 Hz was therefore applied to the circuit of Figure 5.11, as shown in the channel 1 trace in Figure 5.12. The circuit is non-inverting, so an inverting amplifier

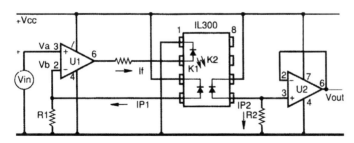

Figure 5.10 *Typical application circuit for the IL300 linear optocoupler, in positive-going unipolar photoconductive model. Although K_1 and K_2 vary with temperature, their ratio K_3 is virtually temperature independent. Devices are coded into bands according to the spreads of K_1 and K_3*

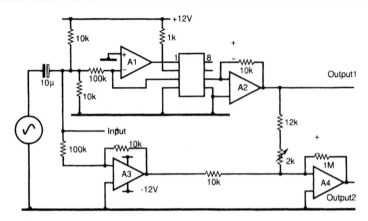

Figure 5.11 *Test circuit used for evaluating the IL300 operating in positive unipolar photovoltaic mode. Ideally, there should be zero resultant signal at output 2*

A_3 was included in the sidechain (input-signal-sample) path, to permit outphasing. After carefully adjusting the 2 KΩ potentiometer to cancel the component in the output which represented the input, the resultant distortion (measured at output 2) is seen to be about 300 μV pp, allowing for the 40 dB gain in A_4. This compares with a wanted signal at output 1 of about 500 mV pp, allowing for the 'gain' of one-tenth from input to output 1. Thus the distortion – assuming it all occurs in the optocoupler with no contribution from the opamps – is well over 60 dB down and is visibly almost pure second harmonic, such as would be expected from a device operated in single-ended mode. (Note that to use this outphasing test method, the ground rails of the input and output circuits have been commoned, whereas of course in practice they would be totally separate – this being the whole purpose of an optocoupler.)

The test was repeated with a 200 Hz input, but this resulted in a large fundamental component at output 2, which could not be outphased. This is

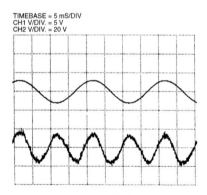

Figure 5.12 *5 V pp 50 Hz input test signal to the circuit of Figure 5.11 (upper trace) and outphased distortion products (lower trace)*

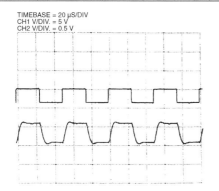

Figure 5.13 *Input and output 1 (Figure 5.11) with a 3.5 V pp 20 kHz squarewave input*

due to the phaseshift via the optocoupler path exceeding that through the outphasing side path, which contains only one opamp as against two and the optocoupler for the signal path. It needs only a twentieth of a degree more phase shift through one path than the other to result in a quadrature component 60 dB down. It cannot be outphased by the potentiometer and is one of the limits to this technique mentioned above. (Adding a balancing delay in the side path – a sniff of CR – would permit complete outphasing of the test signal, provided all frequency components of the test signal were delayed equally; this is clearly easier to arrange with a sinewave test signal consisting of just the one frequency component). To get some idea of the bandwidth available in the photovoltaic mode, a 20 kHz 3.5 V pp squarewave was applied, the input and output 1 waveforms being shown in Figure 5.13. To control ringing, a 10 pF capacitor was added in parallel with the 10 kΩ feedback resistor of A_2 in Figure 5.11. The result agrees well with the 50 or 60 kHz bandwidth quoted by the manufacturer and shown in Figure 5.14(a).

If the two photodiodes and the LED are reversed, the latter being returned to ground rather than $+V_{cc}$, a negative-going unipolar photo-voltaic isolation amplifier results. A bipolar photovoltaic

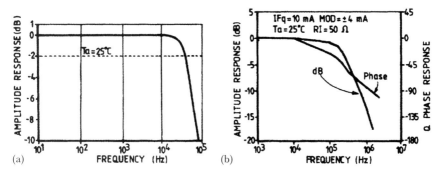

Figure 5.14 *Bandwidth of the IL300 optocoupler: (a) in photovoltaic mode; (b) in photoconductive mode*

amplifier can be constructed using two *IL300*s, with each detector and LED connected in antiparallel. This arrangement provides very low offset drift and exceedingly good linearity, but crossover distortion due to charge shortage in the photodiodes severely limits the bandwidth. Using matched K_3s, with a bipolar input signal centred on ground and taking a hefty 5% as the acceptable distortion limit, the bandwidth is typically less than 1 kHz.

Alternatively, bipolar operation with around 50 kHz bandwidth can be achieved in the circuit of Figure 5.11 by using constant current sources to prebias the amplifier to the middle of its range. A source of zero drift in all the optocoupler circuits discussed here is internal warming of the opamp driving current through the LED, but this can be reduced by using an emitter follower at the opamp's output to drive the LED, shifting most of the dissipation out of the opamp. However, in circuits using prebias, zero drift is also critically dependent upon the quality and stability of the current sources. This being so, one might elect to use the photoconductive mode with its bandwidth in the range of 100–150 kHz (Figure 5.14(b)).

Figure 5.15 shows a bipolar photoconductive isolation amplifier, using rudimentary constant voltage sources for prebias. Note that IP_2 flows through a 60 kΩ resistor against 30 kΩ for IP_1, to restore the gain to unity, allowing for the 2:1 attenuation pad at the input; consequently, twice the prebias voltage is needed in the output circuit. This circuit was substituted for the A_1 and A_2 circuit in Figure 5.11 and the 50 Hz distortion test repeated. (Because the circuit in Figure 5.15 circuit is inverting, amplifier A_3 in Figure 5.11 was not needed and was therefore bypassed.) This time, the amplitude of the 50 Hz input was only 4 V pp, yet the amplitude of the residual was as large if not larger than Figure 5.12: further, its distinct triangularity

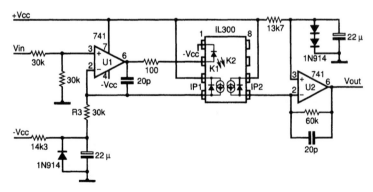

Figure 5.15 *Bipolar (prebiased) photoconductive isolation amplifier*

indicated the presence of significant higher order distortion terms. This illustrates the slightly poorer linearity of the optocoupler in the photoconductive mode. Clearly also, zero drift will be dependent upon the quality of the bias sources, which in Figure 5.15 is not very good. Better performance can be expected from a circuit using devices such as the *LM313*, while an even more ingenious approach is to use a second *IL300* to provide an input circuit with an offset voltage tracking that in the output circuit (Figure 5.16).

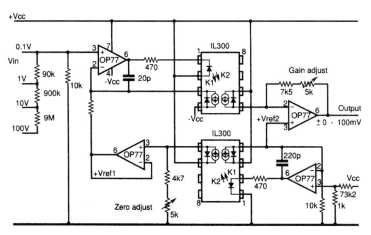

Figure 5.16 *Bipolar photoconductive isolation amplifier using an additional optocoupler to convey to the input amplifier the same prebias voltage used in the output amplifier*

Whether unipolar or bipolar, all the circuits discussed so far have been single ended, i.e. accepting an input which is unbalanced with respect to the input circuit ground. In this case, the CMRR (common mode rejection versus frequency) achieved is simply that provided by the optocoupler itself. In the case of the *IL300*, this is typically 130 dB at 50 Hz falling linearly (in terms of dB versus log frequency) to about 60 dB at 100 kHz. Where the signal source is balanced with respect to the input circuit ground, a much greater CMRR can be achieved using a differential isolation amplifier. The additional isolation comes from the bridge connection of the amplifier on the output side, which combines the inverting and non-inverting inputs to provide a single ended output. Siemens has published differential input circuits operating in both photovoltaic and photoconductive modes, the former offering a bandwidth of 50 kHz combined with a CMRR at 10 kHz of 140 dB (see Reference).

Bin sorting and categories

K_1 (servo gain) is sorted into two bins, each in 2:1 ratios:

Bin W = 0.0036–0.0072
Bin X = 0.0055–0.0110

K_1 is tested at I_f = 10 mA, V_{det} = –15 V. K_3 (transfer gain) is sorted into bins that are ±5%, as follows:

Bin A = 0.560–0.623
Bin B = 0.623–0.693
Bin C = 0.693–0.769
Bin D = 0.769–0.855
Bin E = 0.855–0.950
Bin F = 0.950–1.056
Bin G = 1.056–1.175
Bin H = 1.175–1.304
Bin I = 1.304–1.449
Bin J = 1.449–1.610

$K_3 = K_2/K_1$. K_3 is tested at I_f = 10 mA. V_{det} = –15 V.

The twenty bin categories are a combination of bin sortings and indicated as a two alpha character code. The first character specifies K_1 bins, the second K_3 bins. For example, a code WF specifies a K_1 range of 0.0036–0.0072 and a K_3 range of 0.950–1.056.

K_1, K_3	K_1, K_3
WA	XA
WB	XB
WC	XC
WD	XD
WE	XE
WF	XF
WG	XG
WH	XH
WI	XI
WJ	XJ

The *IL300* is shipped in tubes of 50 each. Each tube contains one category of K_1 and K_3. The category of the parts in the tube is marked both on the tube and on each part.

Acknowledgements

Several of the illustrations in this article are reproduced by courtesy of Siemens plc, Electronic Components Division.

Reference

Designing Linear Amplifiers Using the IL300 Optocoupler, Siemens Appnote 50, March 1991.

> ### Developments in opto-electronics
>
> Opto-electronic ICs have been developing steadily over the years, a trend which will doubtless continue. Some of those in the extensive Texas Instruments range are reviewed in this article, which I 'ghosted' for the name under which it originally appeared.

Light update

Introduction

With the predominance of digital systems in measurement and control applications, comes the increased importance of analog-to-digital conversion, in order to interface real-world (analog) signals to the system. Light is such a real-world signal that is often measured either directly or used as an indicator of some other quantity. Most light-sensing elements convert light to an analog signal in the form of a current or voltage, which must be further amplified and converted to a digital signal in order to be useful in such a system. Important considerations in the conversion process are dynamic range, resolution, linearity and noise. In former times, a discrete light sensor was followed by some form of analog signal conditioning circuitry, before being applied to an ADC, which effectively interfaced it to a digital system. Now, a wide range of intelligent opto sensors are available, combining sensor and signal conditioning in a single device. Typical of these are light-to-voltage converters and light-to-frequency converters.

Light-to-voltage converters

Good examples of these are the *TSL25x* range of single-supply visible-light sensors (Figure 5.17), which combine a photodiode and an opamp connected as a transresistance amplifier, complete with frequency compensation for stability. The photodiode is used without reverse bias, and operates into a virtual earth. There is thus negligible voltage across the diode, minimising dark current. Figure 5.18(a) shows the sensitivity of the three members of the family to

206   Analog circuits cookbook

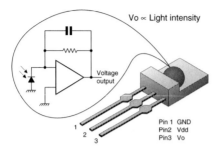

Figure 5.17 *An integrated photodiode plus opamp light to voltage sensor*

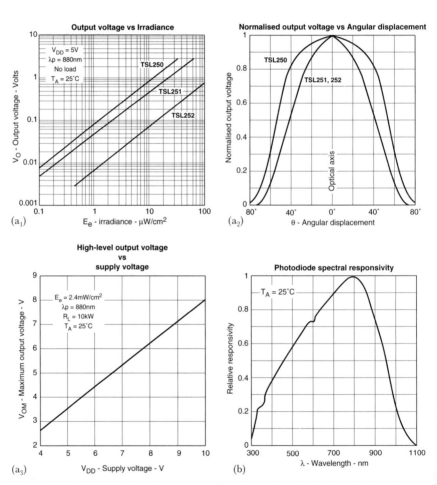

Figure 5.18 *(a) Output voltage as a function of incident illumination for the TSL25x series devices, top, with curves for maximum output against supply, bottom left, and spectral responsivity, right. (b) Angular response of the TSL25x series devices*

illumination on the optical axis, and Figure 5.18(b) shows the relative sensitivity as a function of angular displacement from it. A feature of the *TSL25x* family is a very low temperature coefficient of output voltage V_o, typically 1 mV/°C. This is because the internal feedback resistor (16M, 8M or 2M for the *-250*, *-251* or *-252*) is polycrystalline silicon, with a temperature coefficient which compensates the tempco of the photodiode.

The *TSL26x* range of sensors designed for infra-red applications share the same package and circuit arrangement, and Figure 5.19(a)

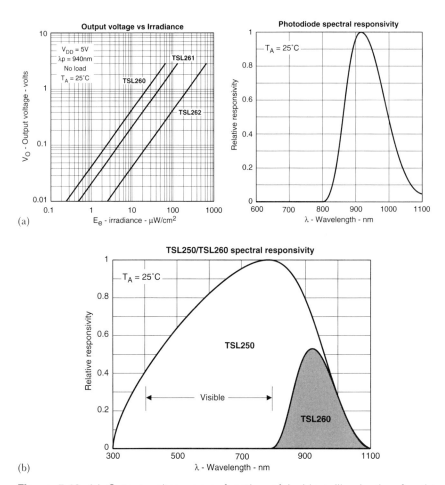

Figure 5.19 *(a) Output voltage as a function of incident illumination for the* TSL26x *series devices, left, together with spectral response, right. (b) Comparing the spectral response of* TSL25x *and* TSL26x *series devices*

shows the on-axis sensitivity of the three members of the family – the angular displacement response is as Figure 5.18(b). Figure 5.19(b) compares the spectral response of the *TSL250* and *-260* families. The data sheet for the *TSL26x* range of devices gives a selection of useful application circuits, which are equally applicable to the *TSL25x* family, see Ref. 1.

Light-to-frequency converters

The light-to-frequency converter is a natural solution to the problem of light intensity conversion and measurement, providing many benefits over other techniques. Light intensity can vary over many orders of magnitude, thus complicating the problem of maintaining resolution and signal-to-noise ratio over a wide input range. Converting the light intensity to a frequency overcomes limitations imposed on dynamic range by supply voltage, noise, and A/D resolution. Since the conversion is performed on chip, effects of external interference such as noise and leakage currents are minimised, and the resulting noise immune frequency output is easily transmitted even from remote locations to other parts of the system. Being a serial form of data, interface requirements can be minimised to a single microcontroller port, counter input or interrupt line, saving the cost of an ADC. Isolation is easily accomplished with optical couplers or transformers. The conversion process is completed by counting the frequency to the desired resolution, or period timing may be used for faster data acquisition. Integration of the signal can be performed in order to eliminate low frequency (such as 50 or 60 Hz) interference, or to measure long-term exposure.

The *TSL220* is a high sensitivity high resolution single-supply light-to-frequency converter with a 118 dB dynamic range, and a convenient CMOS compatible output, in a clear plastic 8 pin DIL package. Figure 5.20(a) shows a block diagram of the internal workings of the device; see also Ref. 2. The output pulse width is determined by a single external capacitor, and the frequency of the output pulse train determined by the capacitor and the incident light intensity, as in Figure 5.20(b). Figure 5.20(c) shows the output frequency as a function of the ambient temperature, normalised to that at 25°C, indicating a need for compensation which can be easily looked after in the subsequent DSP, with the aid of a temperature sensor. The spectral response of the device is very similar to that of the *TSL25X* range shown in Figure 5.19(b), extending a little further into the IR but not quite so far into the UV.

The *TSL235* and *-245* are visible light and IR sensors, packaged in the same 3 pin encapsulations as the *TSL25x* and *26x* ranges, but

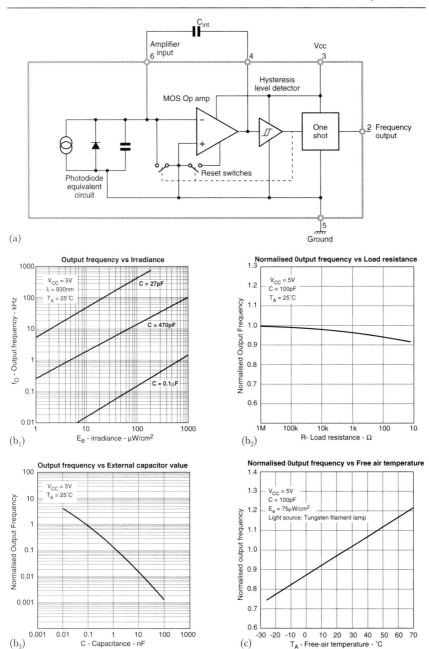

Figure 5.20 (a) Internal workings of the TSL220. (b) Output frequency of the TSL220 versus incident illumination for various values of capacitor, top left, with load and normalised capacitance curves. (c) Output frequency versus temperature, normalised to 25°C, of the TSL220 under the stated conditions

210 Analog circuits cookbook

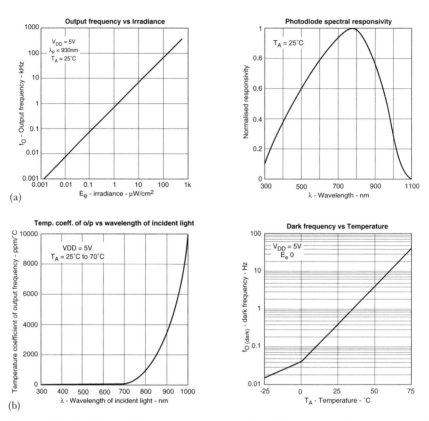

Figure 5.21 *(a) Output frequency versus incident illumination, left, and spectral response, right, for the TSL235 light-to-frequency converter. (b) Temperature coefficient of output frequency of the TSL235, as a function of wavelength, left, and dark frequency performance, right*

producing a frequency output in place of a voltage output. Figure 5.21(a) shows the output frequency versus incident illumination for the *TSL235*, under the stated conditions. Figure 5.21(b) shows how the tempco of output frequency varies with the wavelength of the incident radiation. Note the very low tempco at wavelengths shorter than 700 nm. The *TSL245* is basically the same device as the -*235*, but packaged in an encapsulation material which is transparent in the infra-red but opaque to visible light.

The *TSL230* programmable light-to-frequency converter also consists of a monolithic silicon photodiode and a current-to-frequency converter circuit. A simplified internal block diagram of the device is shown in Figure 5.22(a). Figure 5.22(b) shows how the device simplifies interfacing with an associated MCU. Light sensing is

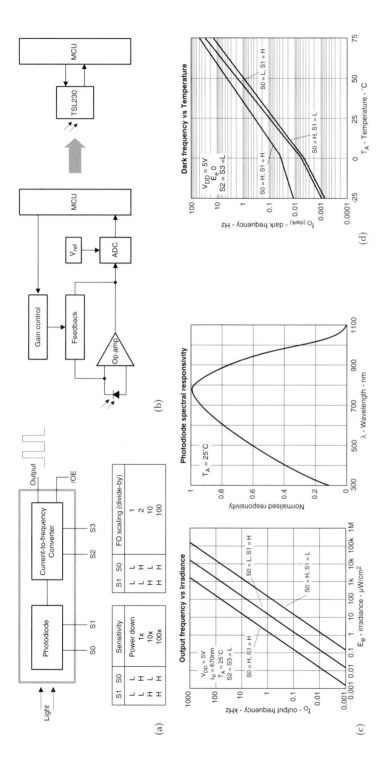

Figure 5.22 (a) Functional block diagram of the TSL230 programmable light-to-frequency converter. (b) Illustrating the system simplification possible with the TSL230 programmable light-to-frequency converter. (c) Illustrating the various sensitivity ranges available to the user with the TSL230, left, together with spectral responsivity, right. (d) Showing the very low dark frequency output of the TSL230, as a function of temperature

accomplished by a 10 by 10 photodiode matrix. The photodiodes, or unit elements, produce photocurrent proportional to incident light. Sensitivity control inputs S_0 and S_1 control a multiplexer which connects either 1, 10, or 100 unit elements thereby adjusting the sensitivity proportionally, implementing a kind of 'electronic iris'. The unit elements are identical and closely matched for accurate scaling between ranges which are illustrated in Figure 5.22(c). The exceedingly low dark current of the photodiode results in the dark frequency output being generally below 1 Hz, Figure 5.22(d).

The current-to-frequency converter utilises a unique switched capacitor charge-metering circuit to convert the photocurrent to a frequency output. The output is a train of pulses which provides the input to the output scaling circuitry, and is directly output from the device in divide by 1 mode. The output scaling can be set via control lines S_2 and S_3 to divide the converter frequency by 2, 10, or 100, resulting in a 50:50 mark/space ratio squarewave.

The *TSL230* is designed for direct interfacing to a logic level input and includes circuitry in the output stage to limit pulse rise- and falltimes, thus lowering electromagnetic radiation. Where lines longer than 100 cm must be driven, a buffer or line driver is recommended. An active low output enable line ($\overline{OE}$) is provided which, when high, places the output in a high-impedance state. This can be used when several *TSL230* or other devices are sharing a common output line.

Like other light-to-frequency converters, the *TSL230* is easily interfaced to digital control systems, but with the added advantage of sensitivity and output frequency range adjustable over a four wire bus, S_0–S_3. Details of interfacing to a particular controller were given in a recent article in *Electronics World*, Ref. 3, but the device interfaces simply with any controller, such as the Texas Instruments *TMS370C010*, the Microchip Technology *PIC16C54HS*, or the Motorola *MC68HC11A8*, see Ref. 1.

References

1. Texas Instruments Intelligent Opto Sensor Data Book.
2. Ogden, F. (1993) An easier route to light measurement. *Electronics World + Wireless World*, June, pp. 490, 491.
3. Kuhnel, C. (1996) Bits of light. *Electronics World*, January, pp. 68, 69.

> This article examines various light sources, mainly LEDs (light emitting diodes), but also some fluorescent fittings. For the LEDs, various drive circuits were derived, and to view the resultant light output, a versatile wideband light meter was developed.

A look at light

Introduction

Lighting emitting diodes have improved enormously, in both efficiency and brightness, over the years. I recall obtaining a sample of one of the first LEDs – red, of course – to become available, in the early 1970s (or was it the late 1960s?). This Texas Instruments device came in a single lead can, with glass window, smaller than TO18, the can itself being the other lead. It was a great novelty to see a wee red light, albeit rather dim, coming out of a solid, but as a replacement for a conventional panel indicator lamp it was really far too dim.

Since then, TI has continued to be a major force in opto products, several of these having been featured in articles in *Electronics World* – Refs 1, 2, 3. But many other manufacturers are active in the field, which covers not only LEDs, photodiodes and phototransistors, but optocouplers, laser diodes, FDDI (fibre optic digital data interface) products and other devices as well. LEDs in particular have seen major advances recently, and being fortunate enough to obtain samples of a number of the latest types, I was interested in finding out just what they will do, and exploring ways of applying them.

Applications a-plenty

LEDs are available covering the whole spectrum, from IR (infra-red) to blue, and have a variety of uses. IR types are used (commonly in conjunction with a photodiode fitted with a filter blocking visible light) in TV remote controls, and in IR beam intruder detectors, etc. High intensity red LEDs are now commonly employed as cycle rear lights, in place of small incandescent filament lamps. They are also suitable as rear lights for vehicles, whilst high intensity amber LEDs are used as turn indicators or 'flashers'. Blue LEDs were for long unavailable, and when they did appear were much less right than devices of other colours. But now, really bright blue LEDs are in production, with a typical application being as one of the primary colours in large colour advertising displays. A good example is the Panasonic *LNG992CF9* blue LED in a T1 3/4 package (surface mount

214 *Analog circuits cookbook*

types are also available). It provides a typical brightness of 1400 mcd over a ±7.5° angle, at a modest forward current of 20 mA.

Whilst most LEDs produce incoherent light, covering a range of wavelengths around the predominant frequency, special types operate as lasers, producing essentially monochromatic light. The result is a beam with very low dispersion, and uses include laser pointers as aids to visual presentations, and as read (and write) sources in optical disk products. Panasonic produce laser diodes also, but these are not at present marketed in the UK, as they are intended for use in consumer products and so available only in large production quantities. Alas, there seems to be no manufacturer of CD players in this country.

Measurements a must

In any branch of engineering – or science in general – little if any progress can be made without suitable measuring instruments. So for my experiments with opto, a lightmeter – with the widest bandwidth possible – was needed. But high sensitivity was equally desirable, and these two parameters face one with an inevitable trade-off. In the event, a medium area silicon photocell was used, operated with zero reverse bias to achieve a low dark current and good noise figure, at the expense of sensitivity.

The circuit design finished up as shown in Figure 5.23, offering a wide range of sensitivities, the sensitivity on range 1 being one hundred thousand times that on range 6. The photocell used was an 'unfiltered' example of the *SMP600G-EJ* (i.e. fitted with a clear window), a sample of which was kindly supplied by the manufacturer (Ref. 4). This is a silicon diode with an area of 4 mm × 4 mm overall, an effective active area of 14.74 mm^2 and a capacitance at zero volts reverse bias of 190 pF. (A rather similar alternative would be *RS 194-076*.) The responsivity as a function of wavelength is as shown by the unfiltered curve in Figure 5.24. The diode is connected to the virtual earth of an opamp, used as a 'transimpedance amplifier'; that is to say, the photodiode output current is balanced by the current through the feedback resistor, giving a volts-out per microamp-in determined by the value of R_f.

The opamp selected is perhaps an unusual choice, but it offers very wideband operation. It has a very low value of input bias current (2 pA typical), although at 20 nV per root hertz, the input noise is not quite as low as some other opamps, especially bearing in mind that the noise is specified at 1 MHz. The $1/f$ voltage noise corner frequency and the current noise are not specified on the data sheet. The *TSH31* has a slew rate of 300 V/μs and a gain bandwidth product

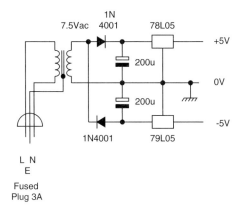
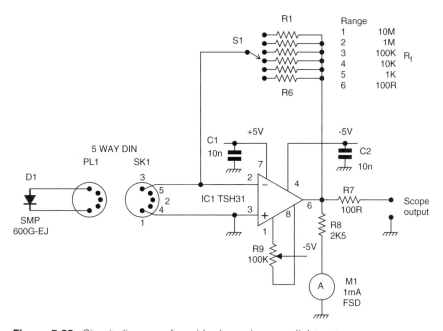

Figure 5.23 *Circuit diagram of a wide dynamic range lightmeter*

of 280 MHz. Given the device's modest open loop gain of ×800 typical, this means that for the higher values of feedback resistor in Figure 5.23, all of the loop gain is safely rolled off by the CR consisting of R_f and the capacitance of the diode, before the loop phase shift reaches 180°. Even on range 6, where R_f is 100 Ω, the circuit is stable – at least with the diode connected. With it removed, the circuit oscillated gently at 160 MHz, so there might be problems if one elected to use

216 Analog circuits cookbook

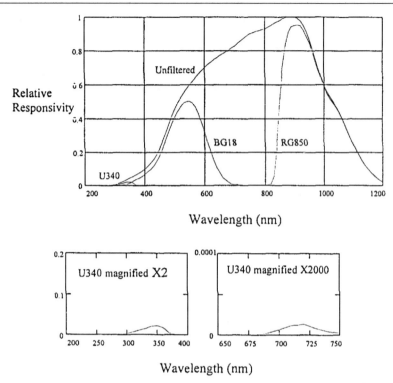

Figure 5.24 *Responsivity as a function of wavelength of the photodiode used in Figure 5.23*

this opamp with a small area diode, having a much lower capacitance. On the other hand, where sensitivity to extremely low light levels (the proverbial black cat in a cellar) is needed, the value of R_f can be raised to 100 MΩ or more, as desired. But note that using a TEE attenuator in the feedback path, to simulate the effect of a very high resistance with more modest values, will incur a severe noise penalty, by raising the 'noise gain' of the circuit. Simply raising R_f instead provides more gain with no penalty of increased noise.

Careful construction was used, with short leads around the opamp and especially for the decoupling components. But for possible further experimentation with different photodiodes, the diode was connected via a 180° five way DIN plug and socket. The board carrying the opamp circuitry was mounted as close as possible to S_1 and the DIN socket. The photodiode was mounted in the backshell of the DIN plug which, being of the better variety with a retaining latch, had a shell of solid metal construction. The rubber cable support sleeve was removed, and the hole reamed out to accept the metal can (a two lead, half height TO39 style) of the photodiode. One lead is connected to the diode's cathode and also to the can, so naturally this lead was earthed. When the diode is illuminated, the anode tries to go positive, and thus sources current which is sunk by the short circuit provided by the opamp's virtual earth. Thus, due to the inverting configuration, the output signal is negative-going.

A small mains transformer with a single 7.5 V secondary winding was used to power the instrument, the opamp being supplied via 78L05 and 79L05 ±5 V regulators. In addition to providing a sample of the opamp output voltage for monitoring on a 'scope, a 1 mA FSD meter was provided. This reads the average value of the photodiode output at frequencies where the inertia of the movement provides sufficient smoothing – i.e. from a few Hz upwards.

Breadboard testing having been satisfactory, the final version was constructed in a small sloping panel instrument case, RS style 508-201. The DIN socket was mounted at the centre back, S_1 top rear, the meter on the sloping panel and the mains transformer as far forward as possible. Provision was made for fitting a screen between the transformer plus power supplies board at the front, and the opamp circuitry at the rear, but in the event this proved unnecessary. Even on the most sensitive range there was no visible hum pickup among the general background noise, which amounted to some 20 mV peak-to-peak on range 1, the most sensitive range.

Measures LEDs and what else

Before getting around to any measurements on LEDs, the instrument was used to check two other sources of light. The first of these made itself felt as soon as the unit was switched on – being the fluorescent light over my laboratory bench. I had gathered the impression that the reason electronic high frequency ballasts produced more efficient lights than tubes operating on 50 Hz with a conventional choke ballast was because the gas plasma didn't have a chance to recombine between successive pulses of current. Whereas the 100 Hz current pulses in a conventional fluorescent fitting with a ballast inductor spend part of their energy re-establishing the plasma each time. (Not that recombination is complete between pulses – if it were, then the starter would need to produce a high voltage kick every half cycle!)

So it was interesting to see the actual variation of light output over a mains cycle, shown in Figure 5.25. This waveform was recorded on range 3 of the lightmeter, with the photodiode head at 50 cm from the tube, a Thorn 2′ 40 W 'white 3500' type – presumably with a colour temperature of 3500°. Given the 5 ms/div. timebase setting, the intensity variations are seen to be, as expected, at 100 per second.

218 *Analog circuits cookbook*

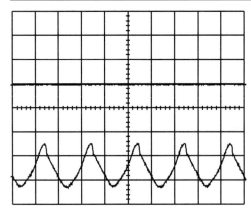

Figure 5.25 *Variation of light output from a 'white' fluorescent lamp. Photodiode at 50 cm from the tube, lightmeter set to range 3. Oscilloscope settings 5 ms/div. horizontal, 0.2 V/div. vertical*

The characteristics of a silicon photodiode, used in voltage (open circuit) mode are non-linear and independent of the diode area. But in current (short circuit) mode, the sensitivity is proportional to the effective area of the diode, and extremely linear versus incident light intensity, over eight or more orders of magnitude, from a lower limit set by the NEP (noise equivalent power) upwards. The zero current line in Figure 5.25, corresponding to complete darkness, is indicated by the trace at one division above the centreline.

So Figure 5.25 shows that between peaks (4.25 divisions below the zero line), the light output falls to just under 60% (2.5 divisions below). There certainly seems to be evidence of a sudden increase of light just after the start of each half cycle of voltage, following the dip. And, of course, being ac, the tube current must go through zero twice every cycle. How brightly the plasma glows at that instant is a moot point, since the light output is mainly due to the tube's phosphors (of assorted colours, to give a whitish light). If the phosphors used have different afterglow times, then there will be variations in 'colour temperature', as well as light output, over the course of each half cycle, just to make things even more complicated.

So I next looked at the radiation from a fluorescent tube without any phosphor, which therefore produced a bluish light. Being entirely without any safety filter, it also produced both soft and hard UV (ultraviolet) radiation. It was a 12" tube type G8T5, used in an electronic ballast powered from 12 V dc. This started life as a camping light, but the original tube was removed and the UV tube fitted when it was converted into a home-made PROM eraser. The unit was fitted into a long box, the front being closed by a removable wide L-shaped PROM carrier. This was to avoid external radiation when in use, as hard UV is bad for the eyes.

With the carrier removed and the photodiode at a distance of 30 cm from the tube, the light output measured on range 3 is indicated by the lower trace in Figure 5.26. The 30 cm separation was more

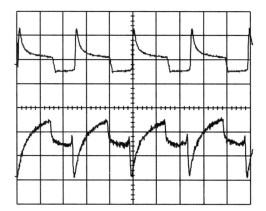

Figure 5.26 *Variation of light output from an uncoated fluorescent lamp. Photodiode at 30 cm from the tube, lightmeter set to range 3. Lower trace: light output as measured by circuit of Figure 5.23, 5 ms/div. horizontal, 0.2 V/div. vertical, 0 V at centreline. Upper trace: waveform of the voltage applied across the tube, via capacitive pick-up, 5 ms/div. horizontal, 2 V/div. vertical, 0 V at two divisions above centreline*

than sufficient to ensure that there was no capacitive coupling between the high voltage waveform applied to the tube, and the photodiode element via the window. This is an important precaution, because the photodiode was not fitted with metal mesh electrostatic screening, available on other models. The upper trace shows the waveform of the voltage applied across the tube. As it was not possible conveniently to get at this directly, it was recorded simply by placing the tip of an oscilloscope probe close to the end of the tube. The waveform at the other end was identical, but of course the other way up. The zero voltage reference for the lower waveform is the graticule centreline. It is clear that the light intensity closely follows the modulus of the voltage waveform, with just a little rounding – which is not in fact due to any limitations of the frequency response of the lightmeter. Presumably this means that the degree of ionisation in the plasma does not vary appreciably over the course of each cycle.

LEDs across the spectrum

It is clear from Figure 5.26 that the electronic ballast ran at a frequency of about 20 kHz, not so very different from a small pocket torch I made a few years back, when the first really bright LEDs appeared. It used a 3000 mcd red LED, powered from a single cell. The circuit is as shown in Figure 5.27, and my records show that the circuit was built and tested as long ago as the end of 1990.

It was housed, along with its AA cell, in one of those small transparent boxes used by semiconductor manufacturers to send out samples – very useful for all sorts of purposes. The typical forward voltage of an LED is between two and three volts, so some kind of inverter is necessary to run it from a single 1.5 V cell. Figure 5.27 uses

220 Analog circuits cookbook

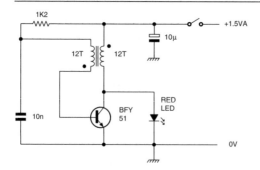

Figure 5.27 *Circuit diagram of a pocket torch using a 3000 mcd red LED*

a blocking oscillator: the resistor provides base current to turn on the transistor and positive feedback causes it to bottom hard. When the collector current reaches a value the base current can no longer support, the collector voltage starts to rise, and positive feedback causes the transistor to cut off abruptly. The collector voltage flies up above the supply rail, being clamped by the forward voltage of the LED. The energy stored in the inductor gives a pulse of current through the LED, which was monitored by temporarily inserting a 1 Ω resistor in its cathode ground return. The current peaked at 150 mA and had fallen to a third or less of this value before the transistor turns on again.

The transformer consisted of a twelve turn collector winding of 0.34 mm ENCU (enamelled copper) wire and a twelve turn feedback winding of 40SWG ENCU, on a Mullard/Philips *FX2754* two hole balun core, which has an A_L of 3500 nH/turns squared. Of course one would not normally expect a 1:1 ratio for a blocking oscillator transformer, but special considerations prevail when designing for such a low supply voltage. The light output is shown in Figure 5.28, measured using range 4 of the light meter, at a range of 1 cm, and the frequency of operation – given the 10 μs/div. timebase setting – can be seen to be a shade under 30 kHz. Although of course of a totally different colour, the red LED torch seemed about as bright as one using a 1.2 V 0.25 A lens-end bulb, whilst drawing, by contrast, only 50 mA. The circuit worked well also with the Panasonic blue LED mentioned earlier.

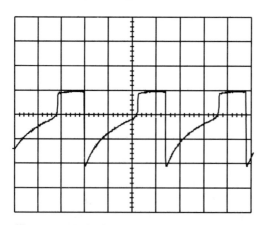

Figure 5.28 *Light output of the circuit of Figure 5.27, measured using range 4 of the light-meter, at a range of 1 cm. 500 mV/div.vertical, 0 V reference line at one division above centreline, 10 μs/div. horizontal*

I recently obtained some samples of very bright LEDs from Hewlett Packard (Components Group), exemplifying the latest technology. The *HLMP-D/Gxxx* 'Sunpower' series are T-1 3/4 (5 mm) precision optical AllnGaP lamps in a choice of red, shades of orange, and amber. These lamps are designed for traffic management, outdoor advertising and automotive applications, and provide a typical on-axis brightness of 9300 mcd.

The *HPWx-xx00* 'Super Flux' LEDs are designed for car exterior lights, large area displays and moving message panels, and backlighting. An *HLMP-DL08*, with its half power viewing angle of ±4°, was compared with an *HPWT-DL00* with a half power viewing angle of ±20°. At a spacing from the photodiode of 1 cm on range 4, with 30 mA in each diode, they gave similar readings, but at greater ranges, the reading from the *HLMP-DL08* exceeded that from the *HPWT-DL00*, on account of its narrower beam. However, the total light output from the *HPWT-DL00* is greater, so it was chosen for an updated version of the LED pocket torch of Figure 5.28.

Brighter still

The resultant circuit was as shown in Figure 5.29, again using an *FX2754* core. Due to its broad beam, the *HPWT-DL00* produced a less bright spot on the opposite wall of the room than a two cell torch with a 2.5 V 300 mA bulb, but only because the latter had the benefit of an extremely effective reflector, giving a very small spot size. With the aid of a small deep curve 'bull's eye' lens (from an old torch of the sort that used a 'No. 8' battery), the Figure 5.29 torch more than held its own, whilst drawing only 150 mA from a single cell. It is thus about four times as efficient as the torch bulb, with a colour rendering that is not so very different – certainly much more acceptable that the red LED torch.

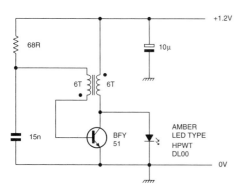

Figure 5.29 *Circuit diagram of the new torch using an* HPWT-DL00 *amber LED, designed to run from a single 1.2 V NICAD cell*

Figure 5.30 shows the performance of the circuit of Figure 5.29. The upper trace shows the collector voltage waveform at 2 V/div. vertical (the 0 V line being at one division above the centreline) and 10 µs/div. horizontal. The lower trace shows the base waveform, also at 2 V/div.

222 Analog circuits cookbook

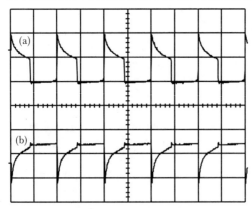

Figure 5.30 *Performance of the circuit of Figure 5.29: (a) collector waveform (upper trace), 2 V/div. vertical, 0 V line at one division above centreline, 10 μs/div. horizontal; (b) base waveform (lower trace), 2 V/div. vertical, 0 V line at two divisions below centreline, 10 μs/div. horizontal*

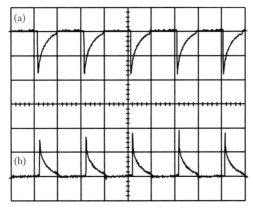

Figure 5.31 *Performance of the circuit of Figure 5.29, continued: (a) lightmeter output (upper trace), 1 V/div. vertical, 0 V line at three divisions above centreline, 10 μs/div. horizontal; (b) diode current waveform (monitored across a 0.18 Ω resistor, lower trace), 50 mV/div. vertical, 0 V line at three divisions below centreline, 10 μs/div. horizontal*

vertical, 0 V line at two divisions below centreline, the operating frequency being about 50 kHz. Figure 5.31 shows the output of the light meter (on range 4, upper trace) at 1 V/div. vertical, 0 V line at three divisions above centreline, 10 μs/div. horizontal, and it is clear that the light pulse has almost completely extinguished by the time that the transistor turns on again to store more energy in the transformer primary. This is seen also in the diode current waveform (monitored across a 0.18 Ω resistor), lower trace at 50 mV/div. vertical, 0 V line at three divisions below centreline.

The peak diode current is just on 400 mA, and although a peak current for the *HPWT-DL00* is not quoted on the data sheet, the average current is safely within the 70 mA maximum allowable at 25°C. The circuit again used a *BFY50* transistor. It also worked with a BC108, although that device actually needed a lower value of base resistor. This was despite its small signal h_{FE} of 500, against the 130 of the *BFY50* – which only goes to show that in a switching circuit, a switching transistor beats one designed for linear applications.

Very bright – but invisible

Figure 5.32 shows the circuit diagram of a little instrument I made up recently for a specific purpose, of which more later. It uses four Siemens infra-red LEDs, type *SFH487*. The unit offers a choice of constant illumination, or pulsed illumination. The three inverter oscillator runs at about 450 Hz, and its output is differentiated by C_4 R_4. This 180 µs time constant, allowing for the effect of R_3 and the internal protection diodes of the inverter input at pin 13 of the *CD4069*, results in a positive-going pulse of about 100 µs duration at pin 8. The string of three inverters speeds up the trailing positive edge of the pulse at pin 13. But if used on their own, a glitch on the trailing edge of the pulse is inevitable, due to internal coupling between the six inverters in IC_1. So C_6 is added to provide a little positive feedback to make the trailing edge of the pulse snap off cleanly.

Figure 5.33 shows the output of the lightmeter when illuminated by the diodes, at a range of 2 cm on range 5. Despite the presence of D_3, there is still some 100 Hz ripple on the supply line. This results in some 100 Hz modulation of the pulse amplitude, and also of the prf (pulse repetition frequency), both visible in Figure 5.33. To show this, a polaroid photograph of the display on a real time analog oscilloscope was used. My simple digital storage 'scope stores only a single trace (per channel) at a time; its facilities do not run to a variable persistence mode such as is found on the more expensive models. Fortunately, for the intended purpose, the 100 Hz modulation was unimportant. The predominant wavelength of the IR radiation from the diodes is 880 nm, this being in the range favoured for physiotherapy purposes. Incidentally, although the spectral bandwidth is quoted as 80 nm, the tail of the spectral distribution evidently extends some way – even just into the visible part of the spectrum – as in operation the diodes exhibit a very feint red glow.

S_1 allows the four IR diodes to be powered by dc, or via Tr_1, with the pulses. Given their aggregate forward voltage of about 5 V, the current through the diodes on CW (dc), determined by $R_{8,9}$ and the supply voltage, is the rated maximum for the devices of 100 mA. In pulse mode, the peak current reaches the rated peak maximum of 1 A. But the duty cycle of approximately 5% keeps the average current to just half of the steady state dc maximum.

The circuit was supplied from an old 6.3 V transformer which was probably intended originally as a TV spare. It would have been used to power the heater of a CRT which had developed a heater/cathode short, thus extending its life and avoiding a costly replacement. This would explain the inclusion of an interwinding screen in such a small,

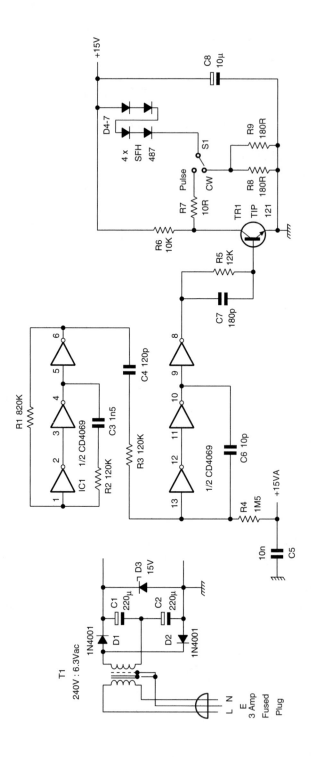

Figure 5.32 Circuit diagram of a high power IR source, with choice of steady or pulsed output

Opto 225

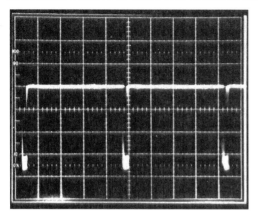

Figure 5.33 *Lightmeter output at a range of 2 cm from the four diodes, on range 5. 1 V/div. vertical, 0 V reference at centreline, 10 μs/div. horizontal*

cheap transformer. In the CW position of S_1, the supply voltage is a shade under 15 V, but tended to rise to nearer 17 V with the lower average current drain in the pulse mode. So D_3 was added to give the designed nominal supply voltage value of 15 V on pulses also. R_6 serves the purely cosmetic purpose of pulling the collector voltage of Tr_1 up to +15 V between pulses. Without it, the voltage lingers at about +10.5 V, since with much less than 5 V across the string of diodes, they become effectively open circuit.

Limitations of the lightmeter

Useful as the lightmeter has proved, it is necessary to bear in mind its limitations when using it. One of these is the sensitivity/bandwidth trade-off mentioned earlier. To illustrate this, Figure 5.34 shows the same waveform as Figure 5.28, the output of the red LED torch of Figure 5.27. But whereas Figure 5.28 was recorded with the lightmeter set to range 4, for Figure 5.34 the light reaching the photodiode was greatly reduced, and range 2 (a hundred times more sensitive) was used. The reduced bandwidth is clearly evidenced by the rounding of the edges of the waveform. With the incident light reduced yet

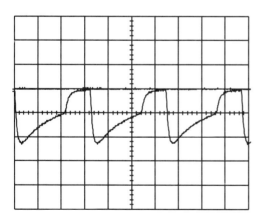

Figure 5.34 *Light output of the circuit of Figure 5.27, measured using range 2 of the lightmeter, at an increased range. 500 mV/div. vertical, 0 V reference line at one division above centreline, 10 μs/div. horizontal*

further and range 1 in use, the waveform was reduced almost to a triangular wave. But while waveform high frequency detail was lost, note that the average value of the incident light is still accurately recorded.

The other great limitation of the lightmeter is, of course, that it provides no absolute measurements. To do so, it would have had to be calibrated with a standard light source, and none was available. Even then, absolute measurements would be difficult, as they always are in photometry. This is especially true when comparing 'white' light sources of different colour temperatures, and even more so with LEDs where typically about 90% of the output radiation is within ±5% or less of the predominant wavelength. Nevertheless, the instrument is exceedingly useful for comparing like with like, and for studying the variations of light output of a source as a function of time.

Its versatility can be further increased by using one of the filtered diodes. Using a diode with the U340 filter (see Figure 5.24), the blue LED tested earlier produced zero response even on range 1. Its predominant wavelength lambda is 450 nm and the spread delta lamba quoted as 70 nm, although the data sheet does not say whether this represents the 50%, 10% or 1% power bandwidth. But evidently there is no significant tail to the distribution extending as far into the UV as 375 nm, where the U340 filter cuts off. But the UV filtered diode did show a small output when held close to a 60 W bulb, due to the very small filter response shown in Figure 5.24, in the region of 720 nm.

Medical uses

I have always been interested in the medical possibilities of electronics, perhaps through having a doctor for a sister. Clearly, though, one should be very wary of experimenting in this area. Some medical applications of optoelectronics are spectacular and hence deservedly well known, such as the use of laser radiation to stitch a detached retina back in place. Other uses are less well known, but one, the use of IR radiation in physiotherapy, I have personal experience of.

It was used, with great success some years ago, to treat supraspinatus tendonitis, alias a painful right shoulder. At the time, an IR laser with just 5 mW output was used, although since then equipments with 50 mW output have become available. The low dispersion offered by a laser source, means that the energy can be applied with pinpoint accuracy to the affected spot, very useful when the power available is low. But I was advised by a physiotherapist (with a degree in physics and an interest in electronics) that apart from this, there is no reason to suppose that an IR laser has any specific advantage over any other source of IR. So having recently experienced a return of the tendonitis, the unit of Figure 5.32 was designed and constructed to treat it. Despite my earlier warning about experimenting, this seemed a safe enough procedure, given that both the condition and the treatment had been previously properly diagnosed.

At 100 mA forward current, the four diodes provide a total radiant flux of 25 mW each – candelas or lumens are inappropriate units for a diode emitting invisible radiation. They were mounted as close together as possible on a scrap of 0.1 inch pitch copper strip board, each angled slightly in so that their beam axes crossed at about 1 cm out. It is thus possible to flood the affected area with IR radiation, where, the theory goes, it 'energises the mitochondria', the chemical power house of each cell, promoting healing. I am happy to report a marked improvement, following a few five minute sessions on alternate days. The pulse mode was incorporated to allow for the possibility that the effect is non-linear with respect to intensity. Instead of half the radiation producing half the effect, and a quarter just a quarter, it might be that half the radiation intensity produced only a tenth of the effect, and a quarter none at all. But the interim conclusion of my limited experience suggests that there is little difference between the efficacy of the pulse and CW modes.

References

1. Hickman, I. (1992) Sensing the position. *EW+WW*, Nov., pp. 955–957.
2. Hickman, I. (1995) Reflections on optoelectronics. *EW+WW*, Nov., pp. 970–974.
3. Robinson, Derek (1996) Light update. *Electronics World* Sept., pp. 675–679.
4. SEMELAB plc, Coventry Road, Lutterworth, Leicestershire LE17 4JB. Tel. 01455 556565, Fax. 01455 552612, Tlx. 341927.

6 Power supplies and devices

> **Battery economy**
>
> Many electronic instruments require portable operation and are therefore powered from internal batteries. This article examines the characteristics of various popular battery types and suggests ways in which their useful service life could be extended. Since the time of writing the relative costs of various cell and battery types have changed considerably, and a much wider choice of primary cells and battery types is now available.

Battery-powered instruments

The use of batteries as the power source for small electronic instruments and equipment is often convenient and sometimes essential. The absence of a trailing mains lead (especially when there is no convenient socket into which to plug it) and the freedom from earth loops and other hum problems offset various obvious disadvantages of battery power. When these and other considerations indicate batteries as the appropriate choice, the next choice to be made is between primary and secondary batteries, i.e. between throw-away and rechargeable types.

Rechargeable versus primary batteries

Rechargeable batteries offer considerable savings in running costs, though the initial cost is high. For example, direct comparisons can be made between certain layer-type batteries, e.g. PP3, PP9, and also certain single cells, e.g. AA, C and D size primary cells, where mechanically interchangeable, rechargeable nickel/cadmium batteries and cells are available. These cost about three to ten times as much as the corresponding zinc/carbon (Leclanché) dry batteries or cell, and in addition there is the cost of a suitable charger. This doubtless accounts for the continued popularity of the common or garden dry battery. Another point to bear in mind is that, contrary to popular belief, the ampere-hour capacity of many nickel/cadmium rechargeable batteries is no greater than (and in the case of multicell types often considerably less than) the corresponding zinc/carbon or alkaline battery. Nevertheless, where equipment is regularly used for long periods out of reach of the mains, rechargeable batteries are often the only sensible power source – a typical example would be a police walkie-talkie. In other cases the choice is less clear; for instance, an instrument drawing 30 to 35 mA at 9 V, and which is used on average for four hours a day five days a week, would obtain a life of 100 hours or more from a PP9 type dry battery (to an end point of 6.5 V, at 20°C).

Assuming the cost of a PP9-sized rechargeable nickel/cadmium battery and charger is 25 times the cost of a PP9 dry battery, it would be two and a quarter years before the continuing cost of dry batteries would exceed the capital costs for the rechargeable battery plus charger. (The effects on the calculation of interest charges on the capital, inflation and the very small cost of mains electricity for recharging have been ignored.)

Using primary batteries

Often, then, the lower initial costs will dictate that a product uses primary batteries, and any measures that can reduce the running costs of equipment so powered must be of interest. When a decision to use primary batteries has been taken, there are still choices to be made, one of which is the choice between layer type batteries and single cells. Sometimes designers prefer to use a number of single cells in series to power a piece of equipment, rather than a layer type battery. The main advantage here is a wider choice of 'battery' voltage by using the appropriate number of cells, although if the usual moulded-plastic battery holders are used, one generally arrives back at a voltage obtainable in the layer type.

The other advantage of using individual cells is that the user then has the choice of primary cells other than zinc/carbon, such as alkaline batteries. On low to medium drains with an intermittent duty cycle, e.g. radio, torch, calculator, these will give up to twice the life of zinc/carbon batteries. However, they are approximately three times the price and therefore the running cost is greater. With very high current requirements and continuous discharge regimes the ratio of capacity realised (alkaline: zinc/carbon) would be increased.

One of the main disadvantages of batteries is that they frequently prove to be flat just when one needs them. As often as not, this is because the instrument has inadvertently been left switched on. If the batteries are of the zinc/carbon type, they can then deteriorate and the resultant leakage of chemicals can make a very nasty and damaging mess. (If the batteries are rechargeable nickel/cadmium types this problem does not arise: modern nickel/cadmium batteries are not damaged by complete exhaustion. However, note that if a nickel/cadmium 'battery' is being assembled from individual cells, they should all be in the same condition – ideally new – and in the same state of charge. Otherwise one cell may become exhausted before the rest and thus be subject to damaging 'reverse charging' by the others.)

A really effective 'on' indicator on a battery-powered instrument might prevent this lamentable waste of batteries. But of the many types of 'on' indicator used, nearly all have proved of very limited effectiveness. One well-known manufacturer uses a rotary on/off switch, the part-transparent skirt of the knob exposing fluorescent orange sectors when in the 'on' position, and this is reasonably effective when the front panel is in bright light. Indicator lamps have also been used but usually with intermittent operation to save current. Examples are a blocking oscillator causing a neon lamp to flash, and a flasher circuit driving an LED. Unfortunately, the power that can be saved by flashing a lamp is very limited. The flashing rate cannot be much less than one per second or it may fail to catch one's attention. On the other hand, the eye integrates over about 100 ms, so flashes much shorter than this must also be much brighter to give the same visibility. Thus a saving of about ten to one in power (ignoring any 'housekeeping' current drawn by the flasher circuit) is about the limit in practice.

I must have ruined as many batteries as most people by inadvertently leaving equipment switched on when not in use, and decided many years ago that the only effective remedy was to replace the on/off switch by an 'on' push button. This switches the instrument on and initiates a period at the end of which the instrument switches itself off again. Clearly it would be most annoying if just at the wrong moment – say when about to take a reading – the instrument or whatever switched itself off, so the push button should also, whenever pushed, extend the operation of the instrument to the full period from that instant. One can thus play safe, if in doubt, by pressing the button again 'just in case'. The period for which the instrument should stay on is, of course, dependent on its use and the inclination of the designer. However, a very short period – a minute or less –

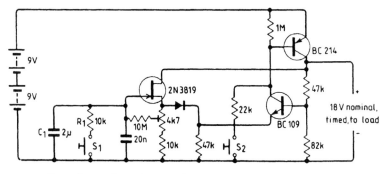

Figure 6.1 *Ten-minute timer designed by the author in 1969*

would generally be rather pointless; provided one had one hand free one would then be better off with a straightforward 'on whilst pressed' button, which is also cheaper and simpler. For many purposes, ten or fifteen minutes is a suitable period, but clearly it is not critical unless the equipment is exceedingly current hungry. After all, it is being left on overnight (or over a weekend) that ruins batteries controlled by an ordinary switch, not the odd half hour or so.

In the late 1960s when I first used a ten-minute timer to save batteries, producing such a long delay economically, and with little cost in 'housekeeping' current was an interesting exercise, especially as monstrously high resistances were ruled out as impractical or at best expensive. So the circuit of Figure 6.1 was developed and proved very effective. The preset potentiometer was set to pick off a voltage just slightly positive with respect to the gate of the *n*-channel depletion FET, so that only a small aiming potential was applied across the 10 MΩ resistor to the timing capacitor, C_1. Thus pressing the 'on' button sets the complementary latch, turning on the instrument and initiating a bootstrapped ramp at the source of the FET. This eventually turned off the latch and hence the instrument, unless the button were pressed again first. In this case the capacitor was discharged again via R_1 and the second pole of the two-pole 'on' button $S_1 + S_2$, and the interval updated. This circuit was very effective in saving batteries, although the exact 'on' period was rather vague due to variation of the gate bias voltage of the FET with temperature. Incidentally, the purpose of the 0.02 µF capacitor was to enable the preset potentiometer to be set for a 6 second period before the 2 µF capacitor was connected in circuit. This made setting up the 600 s 'on' period much less tedious.

An even simpler circuit is possible with the advent of VMOS power FETs, and this is shown in Figure 6.2. The circuit works well in practice, but whilst it might be handy for incorporation in a piece of

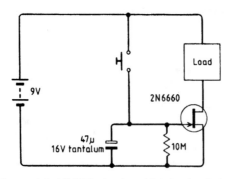

Figure 6.2 *VMOS circuit, which is simple but which does not turn off cleanly*

home-made equipment, it has major drawbacks. Firstly, the data sheet maxima for the FET gate leakage plus that of the tantalum capacitor could result in an 'on' time much less than that predicted by the time constant of 47 µF and 10 MΩ. Secondly, there is no clear turn-off point. As the gate-source voltage falls below +2 V, the drain resistance rises progressively, gradually starving the load current rather than switching it off cleanly. This might be handy if you like your transistor radio to fade out gradually as you go to sleep, but it is not generally a useful feature.

With such a wide choice of integrated circuits available it is possible nowadays to obtain long delays much more easily, and one way is simply to count down from an *RC* oscillator using readily available values of resistance and capacitance. Various timer ICs are available working on this principle, although for a dry battery-powered instrument, where current saving is always a prime consideration, obviously TTL types are less desirable than CMOS. The *CD4060* in particular can form the basis of a timer providing an 'on' interval of up to half an hour with only a 0.1 µF timing capacitor, as in Figure 6.3. Here, on operating the push button, the complementary latch is set, switching on the output, which starts the oscillator with the count at zero. The divide by 2^{14} output at pin 3 is

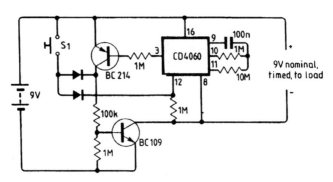

Figure 6.3 *Delay circuit using an oscillator, followed by a counter. Very long delays can be obtained by this method*

therefore at logic 0, holding on the p-n-p transistor and hence the n-p-n transistor in saturation. On reacting a count of 2^{13}, the output at pin 3 rises to the positive rail, turning off the p-n-p transistor and hence the n-p-n transistor and the output. Clearly, by increasing the timing resistor and capacitor at pins 10 and 9 respectively, delays of many hours could be obtained if required.

Such a timing circuit is reasonably cheap to incorporate in an instrument and needs no setting up. As shown in Figure 6.3 it is capable of supplying up to 10 mA or more load current; larger load currents simply require the 100 kΩ resistor in the base circuit of the *BC109c* transistor to be reduced in value as appropriate. The circuit will switch off quite reliably, even though an electrolytic capacitor be fitted in parallel with the load to give a low source impedance at ac. If a DPST push button is used, the circuit can be further simplified by the omission of the two diodes. The small, but nevertheless finite, 'housekeeping' current drawn by the circuit of Figure 6.3 means of course that while 'on', the battery is actually being run down slightly faster than if an on/off switch were used. However, in practice this is more than offset by the reduced running time of the equipment. Quite apart from inadvertent overnight running, an equipment fitted with an automatic switch-off circuit is usually found to clock up considerably fewer running hours during the normal working day than one with a manual on/off switch.

With modern ICs, counting down from an oscillator running at a few Hz is not the only way of obtaining a long delay with modest values of *R* and *C*. Figure 6.4 shows an updated version of the bootstrap timer on Figure 6.1, which could be preferable for use in a sensitive instrument where interference might be caused by the fast edges of the oscillator in Figure 6.3. The analog delayed switch-off

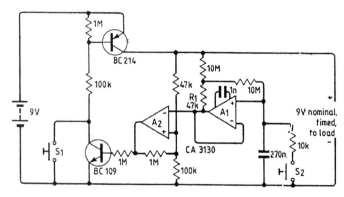

Figure 6.4 *Analog delay circuit avoids possibility of interference from oscillator. A_1 and A_2 are CA3130*

circuit of Figure 6.4 achieves the long delay by applying a very much smaller forcing voltage to the 10 MΩ timing resistance than the reference voltage at the non-inverting input of A_2. With the values and devices shown, no setting up is required as this forcing voltage is still large compared with the maximum offset voltage of the *CA3130*, A_1. For longer delays the 47 kΩ resistor R_1 may be reduced, but to obtain consistent results it would then be necessary to zero the input offset voltage of A_1 (or to use a more modern micropower opamp with an input offset specification in the tens of microvolts region). This circuit will also switch off reliably with an electrolytic bypass capacitor connected across its output.

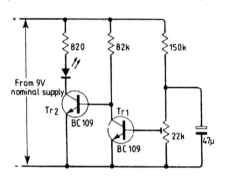

Figure 6.5 *Low battery-voltage indicator. LED illuminates when supply falls below designed minimum*

Figure 6.5 shows a useful and inexpensive battery-voltage monitor which may be connected across the output of either of the circuits of Figures 6.3 and 6.4. The present potentiometer can be set so that the front-panel mounted LED illuminates when the supply voltage falls below the design limit, e.g. 6 V. The temperature coefficient of the voltage at which the LED illuminates is approximately –20 mV/°C, which is generally acceptable, but this can be considerably reduced if required by connecting a germanium diode in series with the lower end of the 22 kΩ preset pot. The 47 µF capacitor delays the build-up of voltage at the base of Tr_1 on switch-on, causing the LED to illuminate for a second or so, assuring the user that batteries are fitted in the instrument and are in good condition. If the voltage falls to an unserviceable level whilst the instrument is on, the LED will illuminate again. The extra current drawn by the LED will cause a further fall in battery voltage, resulting in a sharp, well-defined turn-on. Current drawn while the LED is off is minimal, but by connecting the monitor downstream of a delayed switch-off circuit, even this small current is only drawn whilst the instrument is on.

Choosing the battery size

Using one of the above circuits can reduce the average daily running time of an equipment by a useful amount (as well as eliminating overnight run-down), but the question still remains: 'which dry battery to use?' Circuit design considerations usually dictate the minimum acceptable supply voltage. If a 6 V nominal supply is chosen, a wider choice of capacities is available using four single cells rather than a layer-type battery, but for many purposes an end of life voltage of around 4 V is too inconvenient. A 9 V battery can provide a more useful end of life voltage, whilst if a higher voltage is required, two-layer type batteries in series can be used, 6 V or 9 V types as required.

To decide what size battery of a given voltage to use, refer to the battery manufacturer's data. Tables 6.1 to 6.3 give the total service life in hours to various end voltages (at 20°C) for three different types

Table 6.1 *PP3 estimated service life at 20°C*

Milliamps at 9.0 V	Service life in hours to endpoint voltages of: 6.6 V	6.0 V	5.4 V	4.8 V
Discharge period 30 mins/day				
10	26	29	32	34
15	17	19	21	23
25	9.2	11	12	13
50	2.3	4.1	5.1	5.8
Discharge period 2 hours/day				
1.5	180	190	200	205
2.5	112	122	132	136
5.0	56	62	68	72
10.0	24	28	31	34
15	14	16	20	22
25	–	7.4	9.5	11
Discharge period 4 hours/day				
1	322	355	395	412
1.5	215	240	260	277
2.5	132	147	162	167
5	63	71	77	82
10	17	23	30	33
15	–	12	17	19
Discharge period 12 hours/day				
0.5	695	750	785	815
1.0	365	390	417	427
1.5	240	262	277	292
2.5	125	152	162	170
5.0	44	54	62	72
7.5	18	22	29	36

Note: Also available are the higher capacity PP3P for miniature dictation machines, etc. and the PP3C for calculator service.

Table 6.2 PP6 estimated service life at 20°C

Milliamps at 9.0 V	Service life in hours to endpoint voltage of:			
	6.6 V	6.0 V	5.4 V	4.8 V
Discharge period 4 hours/day				
2.5	492	517	535	545
5.0	240	270	287	302
7.5	142	166	173	194
10	93	111	124	137
15	51	63	73	81
20	33	42	49	55
25	23	30	35	40
50	–	7.8	10	12
Discharge period 12 hours/day				
0.75	2075	2200	2325	2400
1.0	1510	1650	1760	1815
1.5	965	1080	1155	1200
2.5	532	620	635	690
5.0	214	263	202	312
7.5	117	147	109	187
10	75	97	111	127
15	38	50	59	69
25	16	21	25	31
Discharge period 30 mins/day				
50	15	19	22	24
75	0.6	10	12	14
100	25	6.2	7.8	9.1
150	–	2.3	3.5	4.4
Discharge period 2 hours/day				
7.5	169	194	205	215
10	117	140	151	161
15	67	83	93	99
25	31	38	45	48
50	8.9	12	14	16

of layer batteries. The top value PP9 and the ubiquitous PP3 represent the upper and lower capacity ends of the range, whilst the PP6 is one of the three intermediate sizes – PP4, PP6 and PP7 in order of increasing capacity – which, while readily available, are not quite so commonly used. It is important to note that the tables give the service life in hours for the stated current at 9 V with a constant resistance load. Thus the current provided at, for example, an end point of 6 V is only two-thirds of that in the left-hand column of the table.

Table 6.3 PP9 estimated service life at 20°C

Milliamps at 9.0 V	Service life in hours to endpoint voltages of: 6.6 V	6.0 V	5.4 V	4.8 V
Discharge period 30 mins/day				
125	24	35	40	44
150	16	28	33	37
166.67	12	23	29	33
187.5	7.8	19	24	28
250	1.9	9.3	16	19
Discharge period 2 hours/day				
25	193	233	269	286
33.3	150	180	209	223
37.5	122	147	168	180
50	81	99	113	124
62.5	57	71	82	92
75	41	53	62	69
83.33	33	45	53	60
100	20	32	39	44
125	9.8	19	25	29
150	6.1	13	17	20
Discharge period 4 hours/day				
15	332	370	409	437
16.67	291	336	367	394
18.75	266	294	324	349
20	235	273	304	328
25	180	208	234	251
33.33	115	141	158	176
37.5	96	118	134	148
50	59	75	90	101
62.5	37	51	63	72
75	25	35	46	54
83.33	19	30	38	44
100	12	20	27	31
Discharge period 12 hours/day				
15	292	340	379	407
16.67	254	294	321	352
25	127	151	178	206
33.33	73	89	105	134
37.5	58	71	86	110
50	30	40	47	65
62.5	17	24	30	42

The first fact that strikes one is the much greater milliamp-hour capacity of the PP9 than the PP6 and of the PP6 than the PP3, in each case the ratio approaching 6:1. Yet the price differential is (by comparison) tiny. (The PP3 is also available in alkaline technology types, with a capacity over half that of the PP6, but at a premium price.) It would appear therefore at first sight that it must always pay to use the PP9, or at least the largest battery capable of being accommodated within the confines of the instrument case. In general this is true, except in the case of an equipment drawing only a very small current and/or receiving only very occasional use. Under these circumstances, a larger battery would only be partly used before dying of 'shelf life', and a smaller cheaper battery would be a more sensible choice. In fact, if the current drawn is very small – microamps up to a milliamp or so – it is worth considering saving the cost of a switch entirely and letting the equipment run continuously. It is in any case good practice to replace a layer-type battery every year, regardless of how much or how little use it has had, although in a temperature climate it will often remain serviceable for much longer than this. In tropical climates routine replacement after 6 to 9 months is recommended.

The circuits of Figures 6.3 and 6.4, when 'on', apply the full battery voltage to the circuit, except for a 300 mV or so drop due to the collector saturation voltage of the pass transistor. This being so, the load current is likely to be very nearly proportional to the battery terminal voltage, and hence Tables 6.1 to 6.3 are directly applicable. (Strangely, this is the exception rather than the rule; more of which later.) Thus if a 9 V battery is to be used, Tables 6.1 to 6.3 plus those for the PP4 and PP7 will indicate the optimum style of battery, bearing in mind the load current, daily running time and acceptable end voltage. Having chosen the battery type, a graph can be drawn for the appropriate daily usage to permit interpolation between the current values given in the table, giving an accurate estimate of the total serviceable life. Figure 6.6 is an example of such a graph, for the PP9 battery at 20°, with four hours' daily usage, to an end point of 6.5 V. In my experience, the figures quoted in Table 6.1 are conservative, and although there must be some variation from battery to battery, they can safely be taken as minima rather than typical. This view is confirmed by some informal tests which were carried out some years ago by the laboratories of the Finnish PTT in Helsinki.

There is a growing (and welcome) tendency for Japanese and US battery manufacturers to adopt IEC designations for their products rather than using their national or in-house codes, and we can expect UK manufacturers to follow suit in the next year or two.

A final point about using dry cells is a warning that attempts to recharge them are futile and can be dangerous. Fifty years ago a

Leclanché dry cell was built within a substantial zinc canister which acted as mechanical support as well as negative electrode. Using dc with a substantial superimposed ac component, such a cell could be recharged several times, before the canister punctured and the cell dried out. Modern cells contain so little zinc that attempts at recharging are no longer really worthwhile. Recharging will lead to the evolution of gases which a sealed 'leakproof' cell cannot vent and which the cell constituents cannot recombine. In the case of a layer-type battery, the gas evolved forces the layers apart, leading to an open circuit battery.

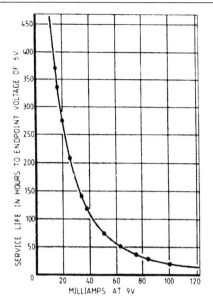

Figure 6.6 *Service life of PP9 battery, used four hours per day*

Stabilised supplies

A piece of electronic test or measuring equipment powered by batteries is often required to possess a degree or accuracy and stability which can only be obtained by operation from a stabilised supply voltage. The current drawn by the instrument at the stabilised voltage is then usually constant, and the data in the tables is thus no longer appropriate. The bulb of a flashlamp likewise tends to be a constant current load, due to its high temperature coefficient of resistance – remember the barretter? On the other hand, the motor of a battery-powered turntable or tape transport with a mechanical or electronic governor tends to draw a constant power, so that the current drawn actually rises as the battery terminal voltage falls. The same applies to stabilisers of the switching variety, which can thus provide a very high efficiency. In practice, for a stabilised voltage of two-thirds of the nominal battery voltage, e.g. 12 V for two PP9s in series, the efficiency of a simple series regulator is almost 66% if the housekeeping current is much lower than the load current, rising to well over 90% at end-of-life battery voltage. This can be held to less than 12.5 V, i.e. an end point of barely over 1 V per cell. The average efficiency of energy usage over the life of the battery is thus over 80%. With rechargeable nickel/cadmium batteries (having an almost constant voltage over their discharge cycle) the figure would be even higher. Whilst a switching regulator can still better this, in a sensitive instrument there can be problems due to the conduction or radiation of interference from the switching regulator into other parts of the circuitry. Thus a supply stabiliser for a battery operated instrument is often likely to be of the conventional series type and the battery current drawn is virtually constant. To estimate the service life of the battery, therefore, tables such as Tables 6.1 to 6.3 cannot be used directly and the following method should be used. For an initial battery voltage E_1, an end-of-life voltage E_2 and a constant current I, the initial load resistance $R_1 = E_1/I$ and the end-of-life resistance $R_2 = E_2/I$. The effective load resistance R_e is defined as $R_e = (R_1 + R_2)/2$ and Figure 6.6 gives battery life (for a PP9), taking $I_e = E_1/R_e$. For dry batteries, since $I_e = 2E_1I/(E_1 + E_2)$, the initial voltage per cell is 1.5 V and the end-of-life voltage 1.0 V, then $I_e = 1.2I$.

An automatic delayed switch-off is just as desirable in a battery-powered instrument incorporating a stabiliser as in one using the 'raw' battery voltage. The circuit of Figure 6.3 incorporates a couple of transistors and it would be elegant and economical to make these function also as the stabiliser circuit. This can be done with just a few extra components as Figure 6.7 shows. Whereas the positive feedback loop of the complementary latch in Figure 6.3 is completed only via the *CD4060* pin 3 output, that in Figure 6.7 is completed independently of the IC. When the zener diode is not conducting, loop feedback is positive and one of the stable states is with both

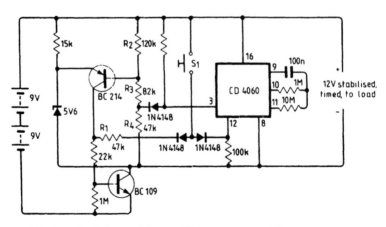

Figure 6.7 *Circuit of Figure 6.3, modified to act as stabiliser*

transistors cut off. Once either transistors starts to conduct, the collector voltage of the *BC109* will fail rapidly until the zener diode conducts, at which point the loop feedback changes from positive to negative and a stable 'on' condition is established. This persists until a count of 2^{13} is reached, when the output of pin 3 of the *CD4060* rises to the positive rail, switching off the p-n-p transistor via the diode. The n-p-n device therefore also cuts off and the 'on' period terminates. The 10 kΩ resistor at pin 3 of the IC is necessary to guarantee the switch-off of the *BC214*, since the p-channel output device of the *CD4060* cannot achieve this unaided when the voltage between pins 8 and 16 falls to a low value.

With the circuit as shown in Figure 6.7, i.e. no load connected, the output voltage will equal the battery voltage whilst the 'on' button is closed. This applies equally at switch-on and when updating the 'on' period. However, for any completed instrument design, once the load current is known it is a simple matter to calculate a value for R_1 which will reliably initiate the circuit without its output exceeding the designed stabilised voltage. In practice also one would provide a preset potentiometer as part of the R_2, R_3, R_4 chain to allow adjustment of the voltage at the base of the *BC214*. This will enable the stabilised output voltage to be set to, say, -12 V exactly, despite the selection tolerance of the zener diode.

As the delayed turn-off circuit of Figure 6.4 also includes a p-n-p and n-p-n transistors, it should be a fairly simple matter to turn these into a stabiliser along the lines of Figure 6.7, though with inverted polarity of course. Such an analog timed stabiliser could be useful where the instrument it powers might be troubled by the switching edges of the oscillator of Figure 6.7. The battery voltage monitor of Figure 6.5 obviously cannot usefully be connected across the output of a stabiliser, nor (although its housekeeping current is only a fraction of a milliamp) would one want to leave it permanently connected across the battery. Figure 6.8 shows how it can be adapted for use with the stabilised delay switch-off circuit of Figure 6.7. The 22 kΩ pot would of course be set to indicate a battery end voltage of 12.5 V.

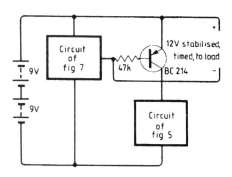

Figure 6.8 *Use of a low-voltage indicator with a stabiliser*

Acknowledgements

Tables 6.1–6.3 are reproduced by kind permission of the Ever Ready Company (Great Britain) Ltd. This company has no connection with Union Carbide, which uses the trademark 'Eveready'.

> ### The MOS controlled thyristor
>
> The MOS controlled thyristor combines many of the advantages of SCRs, high power MOSFETs, GTOs, IGBTs, COMFETs, GEMFETs and MOS-thyristors. The characteristics of this versatile device are explored in this article.

The MOS controlled thyristor

Electronics is often thought of as concerned exclusively with low voltages and currents. Indeed it is often called 'light current electrical engineering' to distinguish it from the heavy currents which are the stock-in-trade of those who deal in megawatts and steam turbo-alternators. But over the years, electronic devices have become big business in the power field, controlling drives in rolling mills, electric locos pulling high-speed trains, etc. In these applications, their function is usually that of a switch: one that does not wear out due to arcing at the contacts, and which can be switched on and off very much faster than any mechanical switch.

    I first came across such devices in the mid-1960s at the Central Research Labs of GEC, when they were still novelties – especially the unencapsulated ones which could be turned on by shining a torch onto the silicon die. At that time the devices occurred in batches of what were supposed to be normal diodes, except that they had been made in a particular much-used silica furnace tube. It was surmised that this contained both p- and n-type contaminants which diffused into the silicon dice at different rates, giving a four-layer structure – as afterwards was shown to be the case. Like all members of the family of thyristors (SCRs, silicon controlled rectifiers) and triacs developed since, these switches could only be turned off by reducing the current through them to zero for long enough for all the minority carriers to recombine; this reinstated the blocking condition, after which they could support a large voltage again without conducting.

    Thyristors (and triacs, which can block or conduct in either direction, making them ideal for ac applications) have developed to the point where they can handle hundreds of amps and volts (Figure

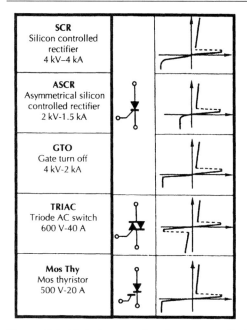

Figure 6.9 *Variations on the silicon controlled rectifier theme. (Reproduced by courtesy of Motorola Inc.)*

6.9), but can need quite a hefty pulse of current to trigger them on. The exceptions are the MOS based devices. One type is similar to an n-channel power MOSFET but with an additional p layer in series with the drain, resulting in a four-layer device. Thus when conducting, the usual FET majority carriers are augmented by the injection of minority carriers, resulting in a lower bottoming voltage. These devices are variously known as COMFETs, GEMFETs, etc., depending on the manufacturer and, like the power MOSFETs from which they are derived, can be turned on or off by means of the gate. Not so the MOS thyristor, which has the usual four-layer structure of an SCR with its very low forward volt drop when conducting, and like them must be turned off by reducing the current through it to zero by external means. However, unlike SCRs, it does not require a sizeable current pulse to turn it on. The GTO (gate turn-off) thyristor can be switched off again by means of the gate, but the drive power needed to do so is considerable.

A recent development has resulted in yet another variation on the thyristor theme, possessing many of the best points of all the various device types mentioned so far. This is the MCT (MOS *Controlled* Thyristor: not to be confused with the MOS thyristor). As Figure 6.10(a) shows, this is basically an SCR, but instead of the base of the n-p-n section being brought out as the gate terminal, the device is controlled by two MOSFETs, one n-channel and one p-channel. These are connected to the anode of the MCT, making it a p-MCT and in effect a 'high side switch'. The p-channel MOSFET can turn the device on by feeding current into the base of the n-p-n section of the complementary latch, whilst the n-channel MOSFET can turn it off again by shorting the base of the p-n-p section to its emitter. To turn the device on, the p-channel MOSFET has only to feed enough

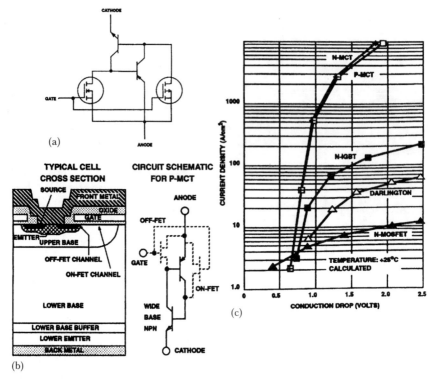

Figure 6.10 *(a) Equivalent circuit of the MCT, showing the complimentary bipolar latch which forms the main current path, the n-channel 'off' MOSFET which shorts the base-emitter junction of the p-n-p section, and the p-channel 'on' MOSFET which feeds base current into the n-p-n section. (b) Cross-section and equivalent circuit of one of the cells of an MCT; there are tens of thousands of these cells in a typical device. (c) Comparison of current capability of the MCT and other devices for a given chip size*

current into the base of the n-p-n section to cause the loop gain of the n-p-n–p-n-p pair to exceed unity, consequently it does not need a very low on resistance. But to turn the device off, the n-channel MOSFET needs to take over the main current, and pass it with a volt-drop lower than the forward V_{be} of the p-n-p section. This description of the operation applies not only to the device as a whole, but also to each and every one of the many thousands of constituent cells (Figure 6.10(b)), so that (if carrying a heavy current) the base-emitter shorting FETs must be turned on uniformly and rapidly to ensure that all MCT cells turn off essentially the same current. If the gate voltage rises slowly, the current will redistribute among the cells, reaching a value in some cells that cannot be turned off.

Power supplies and devices 245

With these devices looking so promising, a data sheet, application note (see References) and some samples were obtained with a view to learning more about them. Taking the simplest possible view, the main current path via the four-layer p-n-p–n-p-n latch should be either on or off, depending upon which of the controlling MOSFETs was last in conduction. An *MCTV75P60E1* in its 5 lead *TO-247* package was therefore connected up as in Figure 6.11(a), the base connections and circuit symbol being shown in Figure 6.11(b). Now the device's input capacitance C_{iss}, that is to say the capacitance looking in at the gate pin with respect to the gate return pin, is listed as the not inconsiderable figure of 10 nF, so as to ensure that the gate received (almost) the full ±18 V pulses which are recommended, the value of C_3 in Figure 6.11(a) was set at 100 nF. When the supplies were switched on, the device did not conduct. Momentarily connecting point X to the −15V rail switched it on, and likewise connecting point X to the +15 V rail switched it off again. The device's 'holding current' (the minimum needed to keep the device in

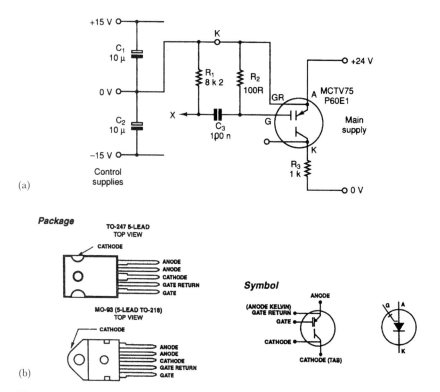

Figure 6.11 (a) Simple on/off test circuit. (b) Base connections and circuit symbols of the Harris MCT

conduction, below which the loop gain falls below unity and the device turns off) is not stated on the data sheet and is merely indicated in the application notes as being 'mA'. With the device switched on, the voltage of the +24 V supply was slowly reduced. At 12 V, the voltage across the 1 kΩ resistor suddenly collapsed to zero, indicating a holding current of 12 mA for this particular sample, at room temperature.

Since the drive was obtained via a capacitor, the drive circuit did not need to be referenced to the gate return pin – this was verified by breaking the circuit at point K and returning the junction of the two 10 μF capacitors to the negative end of the 24 V supply. Thus in certain relatively low power applications, the device could be used as a high side switch without the need for any auxiliary supplies referenced to the high side voltage. It is true that spikes on the main supply could then be coupled to the gate, but due to the large ratio of the 100 Ω gate resistor to the 8K2 recharge resistor, unintentional switching from this cause is not likely, nor is it likely from stray capacitive coupling, given the very large internal gate capacitance. However, this is not the recommended mode of operation, for the following reason. The circuit of Figure 6.11(a) barely tickles the device, given its 600 V blocking capability and 75 A continuous cathode current rating (at +90°C). Therefore the device leakage current was only microamps, way below the current at which the loop gain exceeds unity. Thus the 'off' condition could persist, despite the fact that the bases of the two internal bipolar devices were floating. However, at a case temperature T_c = +150°C, the peak off-state blocking current I_{DRM} (with V_{KA} = –600 V) could be as much as 3 mA, even with the n-channel MOSFET fully enhanced (V_{GA} = +18 V). If the n-channel MOSFET were not fully enhanced, or even off completely, the collector leakage current of the n-p-n bipolar section flowing into the base of the p-n-p section could result in the loop gain exceeding unity; the device would turn on, its blocking ability would have failed. For this reason, the recommended switching and steady state gate voltages are as shown in Figure 6.12.

To meet these requirements, the circuits of Figure 6.13(a) were sketched out, using a *2N5859* (n-p-n) and *2N4406s* (p-n-ps). Both types are switching transistors, rated at 2 A and 1.5 A continuous collector current respectively, so they seemed at first sight a plausible choice since to charge a C_{iss} of 10 nF through 25 V in 200 ns requires just 1.25 A. (Note that the MCT's C_{iss} is relatively constant; it is not augmented during switching by the Miller effect, unlike a power MOSFET.) The circuit was a resounding failure, being quite incapable of swinging the MCT's gate through 25 V in 200 ns. This was presumably due to the fall of current gain of the driver

Power supplies and devices 247

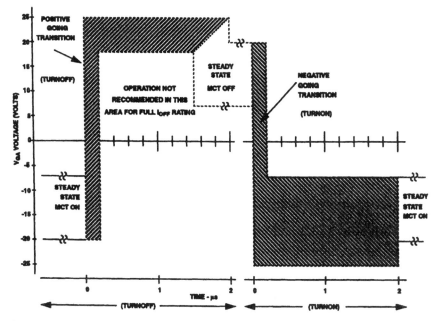

Figure 6.12 *Recommended boundary limits for MCT gate waveform*

transistors with increasing collector current, and the absence of suitable speed-up capacitors. Rather than pursuing the discrete driver approach, therefore, recourse was had to the Unitrode DIL-8 minidip *UC370N* High Speed Power Driver (also available in a 5-pin *TO-220* package) in the circuit of Figure 6.13(b); Figure 6.13(c) shows this device's internal arrangement. Figure 6.14 shows the gate waveform with a 10 kHz TTL squarewave applied to the input of the *UC3705N*, the double exposure showing both positive and negative transitions on MIX timebase (10 V/div. vertical, 20 μs/div. switching to 200 ns/div. horizontal). A 30 V swing across 10 nF results in the 4.5 μJ stored energy being dissipated in the *UC3705N* switch, well within the 20 μJ rating of the n-package and with the 200 ns rise-/falltime in Figure 6.14, the peak current is within the n-package's ±1.5 A peak rating. At 10 kHz the average dissipation is $20\,000 \times 4.5\,\mu J = 90$ mW, again well within the *UC3705N*'s 1 W (25°C) rating.

Note that whilst the *UC3705X* series are specified for operation over the range 0 to +70°C, they incorporate an internal over-temperature shutdown operating at +155°C typical. Shutdown drives the output low, which would turn the MCT on – this will usually be undesirable if not fatal. There are various possible solutions, such as making sure that an external shutdown (perhaps associated with the

248 Analog circuits cookbook

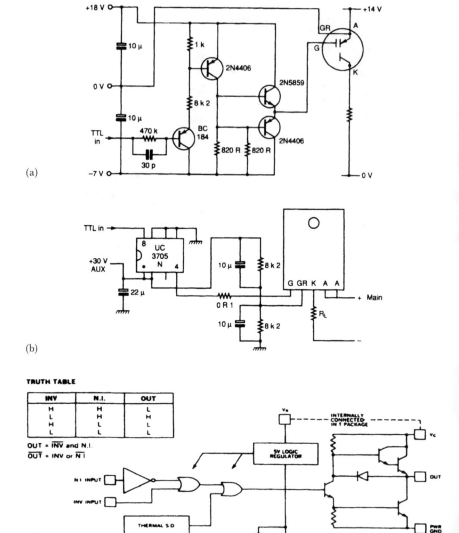

Figure 6.13 *(a) Useless gate driver circuit Mark 1. (b) Gate driver circuit using the Unitrode UC3705N. (c) Internal circuit*

Power supplies and devices 249

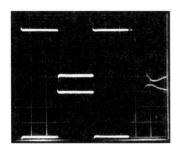

Figure 6.14 *Waveform at MCT gate drive by a 10 kHz squarewave using the UC3705N, double exposure showing both the positive- and negative-going transitions; 10 V/div. vertical, 20 μs/div. switching to 200 ns/div. horizontal*

MCT's heat sink) shuts the whole system down before the UC3705X nears its shutdown limit. An even simpler solution is to use one of the other devices in the UC370XX series such as the UC3706X which has complementary outputs: using the bar (inverted) output will result in shutdown turning the MCT off.

Having ensured that the driver IC circuit was satisfactory, it was time to push the MCT a little nearer its limits. With its 600 V 85 A rating, it is capable of controlling over 50 kW and indeed the manufacturer has produced modules containing 12 paralleled devices with a megawatt capability (Temple *et al.*, 1992). To keep the average power within bounds, the device was pulsed on for 4 μs at a 250 pps rate – a 0.1% duty cycle – and for this purpose the circuit of Figure 6.15(a) was used. Messing about with +600 V on the

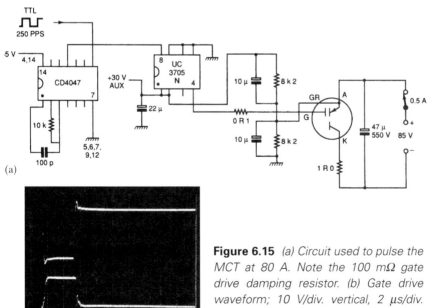

Figure 6.15 *(a) Circuit used to pulse the MCT at 80 A. Note the 100 mΩ gate drive damping resistor. (b) Gate drive waveform; 10 V/div. vertical, 2 μs/div. horizontal (upper trace), voltage across 1 Ω load, 50 V/div. (lower trace)*

250 *Analog circuits cookbook*

lab bench is not a thing to be undertaken lightly, so I settled for a pile of PSUs in series, adding up to a modest +85 V. As these were raw supplies without current limit facilities, a fuse was included for good measure; it blew once on switch-on. This was probably due to the charging current of the 47 µF local decoupling capacitor used across the MCT/load, so after that the mains to the raw supplies was wound up with a Variac. Thereafter, the MCT happily passed pulses of current through the 1 Ω load resistor, the voltage across which is shown in Figure 6.15(b) (lower trace, the upper trace being the gate drive waveform).

My experiments showed that the *MCTX75P60E1* is reliable and easy to use. In applying these devices, one must seek to obtain maximum advantage from their good points, which include a very low forward voltage drop even compared with other minority carrier devices such as IGBTs – let alone MOSFETs – while working within their limitations. As a double injection device – both p- and n-emitters – the MCT conduction drop is well below that of the insulated gate bipolar transistor, especially at high peak currents (Figure 6.16(a)). Clearly their turn-off time will be longer than a

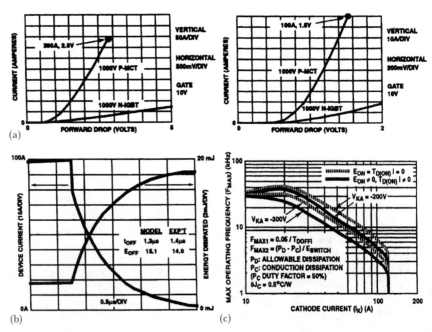

Figure 6.16 Forward conduction drop of the MCT compared with an IGBT. (b) Modelled and actual turn-off losses of 600 V p-MCT (300 V, +150°C inductive turn-off). (c) Maximum operating frequency as a function of cathode current

MOSFET which conducts purely by majority carrier action, although they can be used at higher frequencies than power Darlingtons. Circuit design is eased by the availability of fairly accurate Spice models for the devices; Figure 6.16(b) shows the close agreement between measured and predicted turn-off dissipation, whilst improved Spice models are expected to be available shortly. With the present models, a notional snubber network may be needed to reduce numerical noise in the simulation, but then a snubber may be required for real, depending on the application. This is because the p-MCT's SOA (safe operating area) is rated at half the device's breakdown voltage rather than 80% typical of an n-type power device. If an application involves hard switched inductive turn-off above the SOA and a snubber is not cost-effective, then the MCT is not the best choice. Furthermore, if with a snubber the switching losses now approach the conduction loss, there may be little advantage in using an MCT. On the other hand, with their minimal conduction losses, these devices are ideal in soft switched or resistive load circuits and above all in zero current switched applications such as resonant circuits. The maximum operating frequency F_{max} depends upon both the conduction and switching losses, and can be defined in more than one way (Figure 6.16(c)) (note that 'E' here indicates energy, not emf). From this it will appear that in most applications, the operating frequency will be 30 kHz or lower. A point to bear in mind is that the peak reverse V_{KA} is +5 V, so that in a bridge or half bridge circuit with an inductive load, anti-parallel commutation diodes should be fitted to provide a path for the magnetising current at the start of each half cycle, when operating at low loads.

References

1. *MCT User's Guide*, Harris Semiconductor, Ref. DB307A (contains a list of 39 references to relevant Technical Papers).
2. *MCTV75P60E1, MCTA75P60E1*, Harris Semiconductor, File Number 3374.
3. Temple, V.A.K. *et al.* (1992) *Megawatt MOS Controlled Thyristor for High Voltage Power Circuits,* IEEE PESC, Toledo, Spain, June 29–July 3, 1018–1025 (92CH3163-3).

252 *Analog circuits cookbook*

> ### *Versatile lab bench power supply unit*
>
> For many applications, including audio, a linear lab bench power supply unit (PSU) is preferred over a switcher. For despite its lower efficiency, the linear supply creates no electrical noise – often an essential requirement. This article describes such a PSU, designed not only for excellent regulation and stabilisation, but also to protect any circuitry to which it is connected, in the event of a fault.

Designer's power supply

Introduction

Of the various power supply units (all home-made) gracing my workbench, all are single supplies except one. The exception is a dual 15 V, 1 A unit, with the facility for use as tracking ±15 V supplies, as a 30 V, 1 A supply or a 15 V, 2 A supply. Perhaps because of this versatility, it is the one that gets used most often, despite the fact that the current limit on each section is fixed at 1 A. So it seemed a good idea to start again and design a supply with an adjustable current limit, a design which moreover could be simply varied to give a higher maximum output voltage and/or current, if required – according to whatever mains transformers happened to be available. Since much of my work involves low-level analog signals, in the interests of low noise, the design would be a linear regulator, with the inefficiency that this admittedly involves.

The basics

As the design would spend most of its working life powering circuitry under development, emphasis would be placed upon a generally good performance in the constant voltage (CV) mode (with very low hum ripple even at full load a priority), with performance in constant current (CC) mode somewhat less important. Indeed, the CC mode was intended primarily as a safety feature, to protect both the supply and the circuit under test, in fault conditions. Dual 15 V supplies were envisaged, with provision for independent operation, operation in series and operation with the voltage of one unit (the slave) automatically set to the same value as the other, acting as master. In this mode, the two units may be paralleled to provide double the current available from each separately, or connected in series to provide tracking positive and negative rails.

Specification of the basic 15 V, 1 A Lab Stabilised PSU:

Output voltage:	15 V max. nominal
continuously adjustable	0 V to max. output
noise, hum and ripple	<100 µV rms
Output current:	1 A max. nominal
current limit continuously adjustable	From max. down to 50 µA
Noise, hum and ripple in constant current	<8 mV peak-to-peak

Regulation
Output resistance (not in current limit)	50 mΩ
Peak deviation	700 mV*
Recovery time	10 µs*

* For step load change 50%–100% of rated current

Stabilisation
Output voltage variation	1 mV for ±10% mains voltage change

Mains transformer (for the 15 V, 1 A version):
   rectifier transformer, rated at 21 V dc 1.3 A dc in bridge rectifier/capacitive load service (with 2200 µF reservoir capacitor)

A fairly standard approach, as in Figure 6.17 was adopted, with a CV loop controlled by IC_1 and a CC loop by IC_2. With the wiper of R_v set to ground (fully clockwise), the output voltage is determined by the ratio of R_f and R_i, and the voltage at the NI (non-inverting) input of IC_1. On the other hand, with the wiper of R_v set fully anticlockwise, if R_i/R_f equals R_a/R_b the output voltage will be zero. In CV mode, the CC loop is inactive, since the volt drop across the current sense resistor R_c is small compared with the voltage at the NI input of IC_2.

Of course, Figure 6.17 is purely diagrammatic; in order for it to work, either the opamps must have n-p-n open collector outputs, or the output of each must be connected to the base of the pass transistor via a diode. Furthermore, there must be a dummy load across the stabilised output, to provide a pull-down for the emitter of the pass transistor at low output voltages. But apart from that, the scheme is plausible.

When it comes to the detailed design, practical difficulties emerge. Opamps with open collector outputs are not generally available, and although comparators fill the bill in this respect, they are notoriously unstable when one is so unwise as to try using them in a linear

254 Analog circuits cookbook

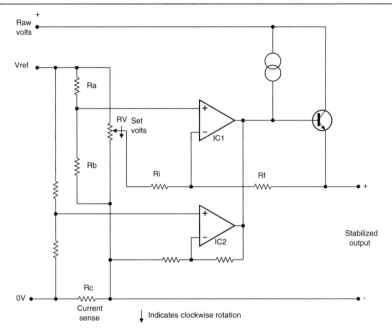

Figure 6.17 *Simplified circuit diagram of a lab bench power supply*

regime. Another problem with the Figure 6.17 scheme is that the opamps must be able to pull the base of the pass transistor right down to the negative stabilised output terminal whilst sinking the current from the constant current generator. But opamps capable of this are limited as to the maximum supply voltage they can stand. So in the event, the 'ICs' in Figure 6.17 were realised with discrete devices. Using discretes provides one with a much greater degree of design flexibility.

The chosen design

This was based upon Figure 6.17, but with a number of variations. For instance, n-p-n current mirrors, such as the Texas Instruments TL0xx range, are readily available, but p-n-p mirrors are not. One could in principle use devices in a pack of matched p-n-p transistors from the RCA CA3xxx range, but the solution adopted here was to use a resistor supplying current from an auxiliary supply of voltage higher than the +raw volts. The final circuit is shown in Figure 6.18. A mains transformer from stock was used, providing a 21 V raw supply, which (allowing for about 2.5 V peak-to-peak ripple across the reservoir capacitor C_3 at 1 A full load) allowed a generous margin of V_{ce} for the pass transistor, even at −10% mains voltage.

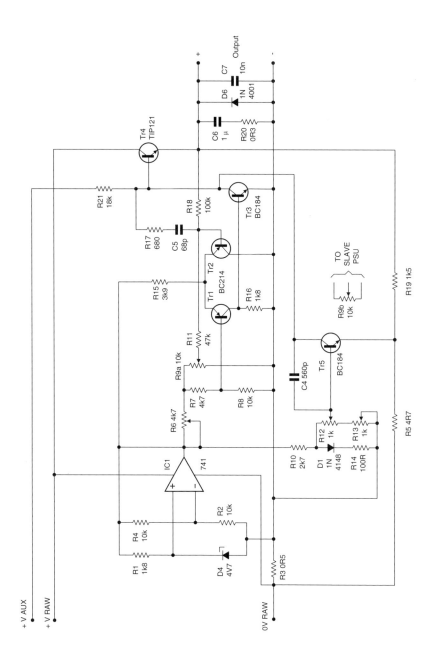

Figure 6.18 Circuit of a 0–15 V power supply with current limit adjustable from 0–1 A

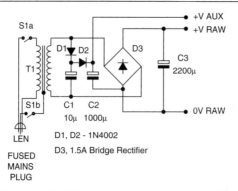

Figure 6.19 *The necessary raw and auxiliary supplies*

The +raw supply, Figure 6.19, uses a bridge rectifier circuit as this makes the best use of the transformer's secondary copper. The modest size reservoir capacitor allows appreciable ripple voltage, resulting in lower copper losses due to a longer conduction angle than would apply with a larger reservoir. An additional half-wave doubler circuit provides the +aux. supply. The reference voltage is provided by an opamp and zener circuit, a convenient arrangement using devices readily available from my component stock, although others may prefer to use their favourite IC voltage reference circuit, of which there are many on the market. The opamp provides the reference for both the CV and the CC loop, and additionally supplies the tail current for the long tailed pair Tr_1 and Tr_2. Together with Tr_3, these do duty as the IC_1 of Figure 6.17. Tr_3 drives the base of the pass transistor, a *TIP121* Darlington device which is adequate for a 15 V 1 A supply, given a generous heat sink. Actually, it is the 18K resistor which drives the pass transistor, Tr_3 simply sinking the excess current as necessary, to maintain the set output voltage. C_6 and C_7 maintain a low output impedance at frequencies where the loop gain is falling off, and in conjunction with these, C_5, R_{17} provide the necessary roll-off of loop gain for the CV loop. R_7, R_8, R_{11} and R_{18} should preferably be 1% metal film, and R_6 permits the CV loop reference voltage to be set to 7.5 V exactly.

In CV operation, Tr_5 remains cut off. At fully clockwise rotation of R_{12} its wiper is at the end of the track connected to R_{13}. This latter is set so that at an output voltage of 15 V, the maximum available output current is, say, 1.1 A. As R_{12} is rotated anticlockwise, the base voltage of Tr_5 is raised, so that a smaller volt drop across R_3 suffices to turn on Tr_5, limiting the available output current to a lower level. Tr_3 and Tr_5 operate as a 'linear OR gate'; whichever pulls the base of Tr_4 lower, that device controls the output voltage.

Unlike the CV loop, the loop gain of the CC loop is quite low, which would result in the short-circuit output current being considerably greater than the maximum current available at an output voltage of 15 V. This undesirable state of affairs is avoided by the judicious application of a little positive feedback from the output. The feedback is applied, via R_{19}, to the emitter of Tr_5, which is returned to the negative end of the raw supply via R_5. Thus as the output voltage falls, the additional drive, necessary to turn on Tr_5 harder, is supplied via its emitter. So an increase in output current, to provide an extra drop across R_3, does not occur. The result is that, with the component values shown, there is actually a small degree of 'foldback', that is to say that the short-circuit current is actually slightly less than the maximum that can be supplied at an output voltage of 15 V.

In addition, $R_{19} + R_5$ form a dummy load, providing the necessary 'pull-down' to enable the output voltage to be adjusted right down to zero. In fact, on no-load, there is a residual output voltage of about 75 mV, even when the demanded voltage is zero. This is due to some 50 µA flowing via R_{11} (whose left-hand end is then at +7.5 V) and R_{18}, producing the said drop across R_{19}. But this residual output voltage is of little consequence since the available current, into a short circuit, is of course no more than 50 µA, even if the CC loop current limit setting be 1 A.

Duals and slaves and meters

The mains transformer used had two similar secondaries, Figure 6.19, and these powered two identical sets of raw and auxiliary supplies (completely isolated from each other) and two almost identical Figure 6.18 type stabiliser circuits. Figure 6.18 actually shows the master supply, R_9 being a two-gang linear 10K potentiometer. R_{9A} controls the output voltage of the master unit. The corresponding 10K pot in the slave is a single gang unit, its track being in parallel with that of the second gang, R_{9B}, of the master unit. In the slave unit, R_{11} is connected to an SPCO switch, which enables the slave's output voltage to be controlled either by its own single-gang R_9, or by the R_{9B} of the master unit. In the latter case, the output voltage of the slave tracks that of the master, enabling their outputs to be paralleled to provide up to 2 A, or connected in series to provide tracking positive and negative supplies.

It is very handy if a power supply has built-in metering, freeing one from the need to wheel up a DVM when setting the output voltage(s). It is particularly convenient when checking a circuit under test for correct operation over the design supply voltage range, such as 4.75–5.25 V. DPMs (digital panel meters) are available at very attractive prices, so built-in metering is no longer a luxury. One popular type is built around the *ICL7106CPL* chip, which is made by a number of semiconductor manufacturers, and such DPMs consist of no more than the IC, an LCD display and a dozen or so discretes. Designed primarily for use in small free-standing DVMs, the IC is usually powered by the ubiquitous 9 V PP3 battery, drawing no more than a miserly 1 mA.

The basic range of a DVM based on this chip is 200 mV, with series limiters and shunts needed for other voltage ranges, and for current ranges. The 200 mV input terminals are designated V_{in} and GD, the input resistance between them being >100 MΩ. However, the CM (common mode) input resistance between these terminals and the negative end of the +9 V supply is undefined. The IC is normally operated with the 9 V battery floating, the GD terminal sitting at about two thirds of the supply voltage, or +6 V. The common mode input resistance, though high, is by no means to be ignored, being non-linear to boot. If the GD terminal is tied to a fixed voltage other than that at which it normally floats, the display shows the overload indication, a lone '1' in the left-hand digit. On the other hand, the need to supply a floating +9 V is clearly an inconvenience for the designer. However, it turns out that with a little ingenuity the 9.4 V reference supply to the CV and CC loops can be pressed into service.

Figure 6.20 shows the scheme: the reference supply is used as a pseudo-floating supply by translating and scaling the 0–15 V output to be measured to a 200 mV range at the 7106's natural common mode input voltage. This is carried out at a high impedance level (possible in view of the DPM's very high input resistance), thus avoiding pulling the common mode input voltage away from its

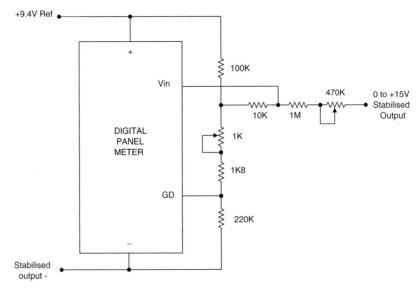

Figure 6.20 *Using the 9.4 V reference supply as a pseudo-floating supply for a DPM*

preferred level. The resistance values required are not what you would calculate on the basis of an infinite common mode input resistance. The proper values are in fact not easily derived, given the non-linear common mode input resistance; they were therefore made up including trim pots, which were adjusted to give the right readings at output voltages of zero and +15 V. As the adjustments interact, they must be iterated to achieve the correct final settings. Adjusted thus, the DPM agreed with the readings on a Philips PM2521 DVM to well within ±1% over 0–15 V range. The latter was reading the actual 0 to +15 V output of the PSU, whilst the DPM saw a 0–150 mV input. But linking the appropriate points on the rear PCB of the DPM, namely jumper P_2, activates a decimal point to indicate a 00.00 to 19.99 range. Three samples of DPM were tested in the circuit of Figure 6.20, only minor readjustments of the trimpots being needed for each.

A second DPM can be used as a dedicated current meter, but an opamp stage would be needed to suitably scale and translate the 0–500 mV developed across R_3 to a suitable level. But my personal preference for a dedicated current meter is a moving coil analog type, since this provides an instantaneous visible indication of the current drawn, and a versatile, fully protected circuit is described later on. Using a DPM, with its reading rate of about three readings a second, and allowing for settling time, no clear indication of the current drawn is instantly available. Indeed, if the current being drawn by the load that the PSU is supplying has an appreciable ripple, the last few digits may be constantly flashing. An analog meter, by contrast, has a degree of built-in smoothing, due to the inertia of the movement. Nevertheless, a digital readout of current can be useful for testing purposes, so perhaps the best of both worlds would be an analog meter permanently indicating the current being supplied, and a DPM normally indicating output voltage, but switchable by means of a biased toggle, to read current when required.

A useful performance

The 15 V, 1 A PSU of Figures 6.18 and 6.19 was tested for the usual performance parameters, with the following results. The dc output resistance measured 50 mΩ, whilst the change in output voltage for a 10% change in mains voltage was barely 1 mV. The output ripple in constant voltage mode, supplying 1 A at 15 V, was estimated at around 200 μV peak-to-peak as measured on the 2 mV/div. range of a Thurlby-Thandar Digital Sampling Adaptor type DSA524 with averaging mode selected. In view of the low signal level, to avoid possible errors due to earth loops, the reading was repeated, using

260 Analog circuits cookbook

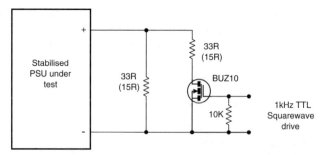

Figure 6.21 *Circuit used for testing the transient response of the PSU*

the AF millivoltmeter section of the Lab-amp described in Ref. 1, with its balanced floating input stage. There was no indication on the 3 mV rms full scale range, confirming that the full load ripple is below 100 microvolts rms. With the same load resistance and set voltage, the current limit was reduced to enter CC mode. The ripple voltage across the load was then 8 mV peak-to-peak at 900 mA (reducing pro rata with current), reflecting the lower gain of the CC loop.

An important parameter of a power supply is the transient response when the demanded load current changes abruptly. Figure 6.21 shows a simple test circuit which was used to switch the load between 0.5 and 1 A approximately, at a rate of 1 kHz. The transient was captured using the DSA524. The result is illustrated in Figure 6.22, at 200 µs/div. (upper trace) with an expanded view of the transient at 5 µs/div. (lower trace). When the load drops from one amp to half an amp, there is a momentary positive-going spike of some 700 mV. But since the width of this measured out at just 100 ns, the energy associated with it is low. Thereafter, there is a well-controlled transient, settling within 10 µs to the steady level. The story when the load switches from 0.5 to 1 A is similar; the spike just looks smaller in the upper trace as a sampling pulse

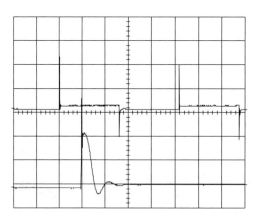

Figure 6.22 *Transient response of the PSU when the load switches between 0.5 A and 1 A: upper trace 200 mV/div., 200 µs/div.; lower trace 200 mV/div., 5 µs/div.*

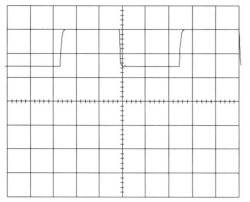

Figure 6.23 Load switching between 33 Ω and 17.5 Ω, with the demanded output voltage set to 15 V but the current limit reduced to roughly 0.5 A, i.e. such that at 17.5 Ω the voltage collapses to 7.5 V: 5 V/div., centreline = 0 V, 200 μs/div.

doesn't happen to have caught the peak. Figure 6.23 shows the same load and set voltage, but with the current limit set to roughly 0.5 A, so that at the lower value of resistance, the output voltage drops to 7.5 V. The response is overshoot-free, as the CC loop is, if anything, overdamped.

The prototype is stable both on- and off-load in both CV and CC modes with 1000 μF in parallel with the output. Of course, a 1000 μF capacitor reduces the 7.5/15 V switching waveform of Figure 6.23 to pretty well an 11 V straight line, and even just 10 μF turns it into something approaching a triangular wave.

Variations on a theme

As mentioned in the Introduction, the circuit is designed to be 'stretchable', both in voltage and current. Typical ratings for commercial lab bench power supplies are 15 V or 30 V, at 1 A, 2 A or occasionally 5 A. Figure 6.24 shows the output of the PSU when the load switches between 1 A and 2 A, the 33 Ω resistors in Figure 6.21 having been replaced by similar wirewound 15 Ω resistors. As the raw supplies, pass transistor Tr_4 and its heat sink were not rated for continuous use at 2 A, the test was not continued for longer than necessary to obtain the results shown.

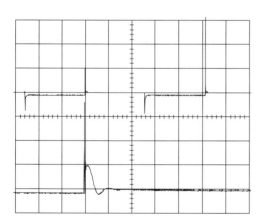

Figure 6.24 Transient response of the PSU when the load switches between 1 A and 2 A: upper trace 500 mV/div., 200 μs/div.; lower trace 500 mV/div., 5 μs/div.

To enable the unit to provide 2 A, even in the short term, the current sensing resistor R_3 was temporarily shorted to defeat the current limit – not a practice to be recommended. A proper 2 A version requires only the beefing up of the raw supplies, a pass transistor with a higher maximum dissipation than the *TIP121* used in Figure 6.18 (with suitable extra heat sinking), and halving the values of R_3 and R_{21}.

Similarly, few changes are required for a 30 V version, other than attention to voltage ratings of capacitors and semiconductors – and one other point. If using a 3.5 digit DPM in a 30 V version, provision must be made to switch the latter from 19.99 V full scale to 199.9 V full scale. A useful halfway house, providing more than 15 V output but without the complication of DPM range switching, is a 20 V design. This will enable circuitry designed for either 15 V or 18 V nominal supplies to be tested at both top and bottom supply limits.

Whatever the rating chosen, a useful feature to incorporate is a non-locking push button wired across the output terminals. Pressing this will put the PSU into current limit, and R_{12} can then be adjusted for a lower limit than the maximum, if required.

More variations

The *TIP121* Darlington is so cheap and convenient, it is worthwhile considering whether it can be used in higher power designs. For example, in a 15 V, 2 A design, two can be used in parallel, each fitted with a 0.5 Ω emitter ballast resistor to prevent current hogging by one of them. The heat sinking must be adequate to cope with the total worst case dissipation (with a short-circuited output) at top mains voltage, but the two devices are equivalent to a single Darlington with half the junction-to-heat sink thermal resistance of a single device.

For even higher powers, the McPherson circuit, Ref. 2, is attractive – the patent is probably by now expired. An updated version of this scheme is shown in Figure 6.25. If you imagine the raw voltage to be only marginally greater than the maximum rated output voltage (e.g. at minimum mains voltage), then only a quarter of the worst case power dissipation ever appears in either transistor, and often much less. For at rated maximum current on short circuit, Tr_1 is cut off, Tr_2 bottomed, and all the dissipation takes place in the ballast resistor R_b ($R_b = V_{\text{rated max}}/I_{\text{rated max}}$). At maximum rated current at maximum output voltage, Tr_2 can make no significant contribution, so all the current is supplied via Tr_1, whose V_{ce} is then, however, minimal.

There are two worst cases; the first is full output current at half output voltage. Here, Tr_2 is bottomed and supplies half the current,

Power supplies and devices 263

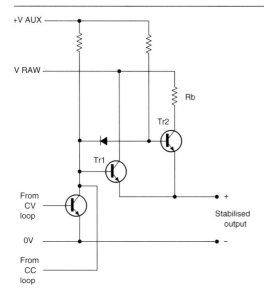

Figure 6.25 An updated version of the McPherson Regulator. Of the worst case total dissipation, only around a third ever appears in either transistor

whilst Tr_1 supplies the other half, with a V_{ce} of half the raw volts. The other is negligible output voltage at half rated current. Here, Tr_1 is off and Tr_2 supplies half the rated current with half the raw volts collector to emitter. Either way, only a quarter of the maximum power dissipation appears in either transistor, and never in both at the same time, so they can usefully share the same heat sink. In practice, the worst case transistor dissipation is somewhat more than this, especially at top mains voltage, but is still much lower than schemes where all the dissipation occurs in pass transistors. Clearly a considerable saving in the heat sinking requirements is achieved. Most of the dissipation occurs in (a) wirewound resistor(s) which can reject heat at a 300°C surface temperature, against 125°C for a semiconductor junction. Ref. 2 describes how the scheme can be extended to four transistors, three with appropriate value resistors in their collector circuits. Turning on one or more as required, in sequence, keeps most of the dissipation in the various ballast resistors, a very effective arrangement.

Variations on the current limit circuit are also possible. Figure 6.28 shows a versatile analog current meter circuit. An opamp is used to amplify the 0.5 V maximum drop across the current sense resistor R_3 to 6.8 V, to drive a 1 mA FSD meter, scaled 0–1 A and 0–300 mA. Other values of feedback resistor may be selected, giving a choice of 30, 100, 300 and 1000 mA ranges. On the most sensitive of these, the full-scale volt drop across R_3 is only 15 mV, so an opamp with low offset voltage is indicated. A *TLC2201/C* being to hand, this device – with its typical offset of 100 µV – was used. In fact, with its low maximum input offset of 500 µV (200 µV on the /AC and /BC versions), the *TLC2201* comes without offset adjust inputs, and at 1 pA its bias current is not large either. But a more mundane opamp, complete with offset adjustment, would suffice. The circuit shown protects the meter against overload. If the PSU supplies 1 A when the meter is switched to the 30 mA range, a 33× overload, the opamp output can only reach something less than +9.4 V, limiting the actual meter overload to less than 50%.

Another variation can be useful where the maximum power available from the raw supply at +7% mains voltage is greater than the pass transistor can dissipate indefinitely with the output short-circuited. For example, on a 15 V 1 A unit, the current limit could be set at 1.5 A at 15 V, folding back to 1 A when the output is shorted – this merely involves raising the value of R_5. A further ploy is to thermally couple Tr_5 to Tr_4; the short-circuit current can then be set to, say, around 1.2 A with the unit cold. On an extended short circuit, the V_{be} of Tr_5 then will fall by about 2.2 mV/°C as the heat sink and pass transistor warm up, gradually reducing the short circuit current back to 1 A.

Tips on using the PSU

With one or two amps available at whatever output voltage has been set, up to 15 or 30 V, there is always the possibility of damage to a newly constructed prototype circuit connected to the PSU, when first powered up. Some engineers are supremely confident of their design and workmanship, and thus have no qualms. For my part, there is always the worry that some misconnection – or even more likely, an undetected solder bridge – will result in the damage or destruction of one or more devices.

A safe way of powering up in such circumstances is to make use of the continuously variable current limit. The PSU is set to the desired output voltage, and the current limit control then set fully anticlockwise, causing the output voltage to collapse to zero. The current meter is then set to a range appropriate to the current which the circuit under test is expected to draw, and circuit under test connected to the PSU. The current limit control can now be advanced slowly clockwise, keeping a weather eye on the current meter and another on the voltmeter. If the current starts to rise alarmingly before the output voltage is anywhere near the preset value, it is prudent to switch off and recheck the circuit under test for faults.

If the PSU is to be used in this way, it is well to use a reliable long-life pot for the current limit control R_{12}, such as a cermet type. There is an alternative mode of use, which though not offering such certain safety, will usually prevent any damage, and is useful where the supply is to be used by all and sundry. This is to fit an ON/OFF switch for the PSU output, independent of the mains ON/OFF switch. Downstream of this switch is a 100 µF capacitor (and discharge resistor) as in

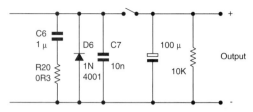

Figure 6.26 *When the separate output switch is closed, the 100 μF capacitor causes the output voltage momentarily to collapse (almost) to zero. The output voltage then ramps up with the PSU in current limit, until the preset voltage is reached, or until the limited available current is drawn through a fault in the circuit under test*

Figure 6.26. On switch-on, charge sharing between C_6 and the 100 μF capacitor will cause the output voltage to collapse to 1% of the preset value, e.g. 15 V down to 150 mV. The output voltage will then ramp up at the set current limit until either the preset output voltage is reached, or the fault current drawn by the circuit equals the current limit. If the latter is only tens of milliamps, more than adequate to power a good deal of CMOS circuitry, usually no permanent damage will result, and the fault can then be cleared at leisure.

Tailpiece

The stratagem described earlier, to permit the DPM to be powered from a non-floating supply, is not always convenient. In this case, an inverter can be used to produce a suitable floating 9 V supply from whatever rail voltage is available. Figure 6.27 shows a very simple flyback inverter for operating a DPM from a +5 V rail. At under 60%, the efficiency when supplying close on 10 V at 1 mA, is not wonderful, but the odd 3.8 mA is hardly a heavy load on the 5 V supply. The

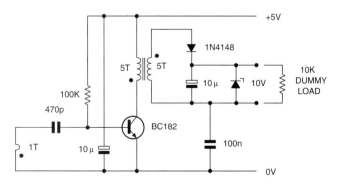

Figure 6.27 *Circuit of a simple flyback inverter, producing a nominal 10 V, 1 mA, suitable for powering a DPM using the ICL7106*

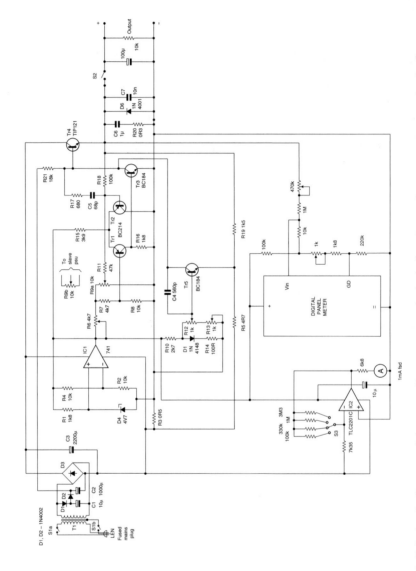

Figure 6.28 Full circuit diagram of the versatile power supply, including the analog current meter circuit. Components shown are for a 15 V, 1 A output, but the circuit is easily altered for other outputs

prototype circuit ran at about 170 kHz, producing 9.52 V off-load, 9.46 V into a 10K dummy load simulating a DPM. The two 5 turn windings were of bifilar wire, on a Mullard/Philips *FX2754* two-hole balun core having an A_L of 3500 nH/turn^2. The output voltage is floating dc-wise, but the 100 nF capacitor is added to prevent switching frequency ripple appearing on the output relative to ground. The circuit is readily adapted for other supply voltages, and as the required output power is less than 10 mW, efficiency will not usually be an important consideration.

References

1. Hickman, I. (1996) Listening for clues. *Electronics World*, July/August, pp. 596–598.
2. McPherson, J.W. (1964) Regulator Elements Using Transistors. *Electronic Engineering*, March, p. 162.

7 RF circuits and techniques

> **Direct conversion FM design**
>
> Although many people are familiar with homodyne (direct conversion) reception of CW and SSB signals, it is not immediately obvious that FM signals can be received by direct conversion. But with some crafty signal processing, the original baseband modulation can indeed be recovered.

Homodyne reception of FM signals

Homodyne or direct conversion reception has always attracted a good deal of attention, especially in amateur circles (Hawker, 1978). It has the attraction of simplicity, both in principle and in hardware terms. Figure 7.1 shows a simple homodyne receiver which could in principle be simplified even further by the omission of the rf amplifier (at the expense of a poorer noise figure) and even of the input tuned circuit or bandpass filter – some filtering might be provided by the aerial, if

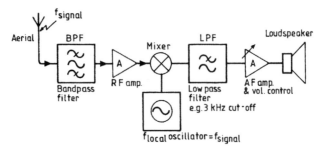

Figure 7.1 *Principle of the homodyne, in which the received signal is converted directly to audio by setting the local-oscillator frequency equal to that of the signal*

for example it were a half wave dipole. The homodyne has something in common with the superhet, but whereas the latter produces a supersonic intermediate frequency (hence SUPERsonic HETerodyne receiver), in the homodyne the local oscillator frequency is the same as the signal's carrier frequency, giving an IF of 0 Hz.

It is well known that a homodyne receiver can be used for the reception of SSB, although in a simple homodyne there is no protection against signals in the unwanted sideband on the other side of the carrier. A small offset between the frequency of the local carrier and that of the SSB suppressed carrier can, however, be tolerated, at least on speech signals. Homodyne reception can also be used for the reception of AM, but no frequency offset is permissible and the phase of the local carrier must be (at least nearly) identical with that of the incoming carrier, otherwise all the modulation 'washes out'. This means in practice that the local oscillator must be phase locked to the carrier of the incoming signal. If the local oscillator is under-coupled so that it barely oscillates, if at all, the incoming signal energy can readily synchronise it, an arrangement universally employed under the name of 'reaction' in the days of battery-powered 'straight' wireless sets using directly heated valves with 2 V filaments.

FSK and the homodyne

CW is readily received by a homodyne, but it is not immediately obvious how it could successfully be employed for FM reception. However, it can, as will shortly become clear. The simplicity of the homodyne means that it is potentially a very economical system and, for this reason, there has always been an active interest in the subject on the part of commercial concerns; a number of homodyne receivers have appeared on the market. Vance and Bidwell (1982) describe a paging receiver which is actually a data receiver using FSK modulation. This is a type of FM where the information is conveyed by changing the signal frequency rather than its amplitude. One could in principle receive the signal by tuning the local carrier just below (or above) the two tones and picking them out with two appropriate audio frequency filters, but this would be a very poor solution, since there would be no protection from unwanted signals on the other side of the carrier.

The solution adopted by Vance and Bidwell was much more elegant, with the local oscillator tuned midway between the two tones, so that both ended up at the same audio frequency, equal to half the separation of the two tones at rf. Now in a simple homodyne receiver this would simply render the two tones indistinguishable; in a practical system it is necessary to have some way of sorting them out. This is entirely feasible, but it does involve just a little more kit than in a simple homodyne receiver of the type shown in Figure 7.1. Before looking at how it is done, some basic theory is needed, which I have chosen to illustrate graphically rather than with algebra and trigonometry, though the results are of course the same.

Figure 7.2(a) shows a sinusoidal waveform and illustrates how its instantaneous value is equal to the projection onto the horizontal axis of a vector of fixed length, rotating (by convention) anticlockwise. Figure 7.2(b) carries the idea a little further and shows two such vectors, representing a sinewave and a cosine wave. As in Figure 7.2(a), both vectors should be imagined as rotating at an angular speed of ω rad/s, that is $(\omega/2\pi)$ Hz. If they really were, they would be a blur at anything much above 10 Hz, so further imagine the paper they are drawn on to be rotating in the opposite direction, i.e. clockwise, at ω rad/s, thus freezing the motion and enabling us to see what is going on. Furthermore, with this convention, one can picture what happens when a slightly different frequency sinewave is also present, say at a frequency of $(\omega + 2\pi)$ rad/s or 1 Hz higher. This can be represented on the vector diagram as a vector rotating anticlockwise at a velocity of 2π radians, or one complete revolution

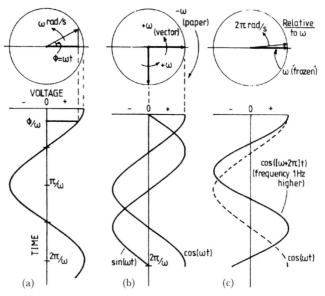

Figure 7.2 Vector representation of sinusoidal waveforms. Waveform at (a) is derived from projection of rotating vector onto horizontal axis, while at (b) two such vectors are shown in quadrature, producing sine and cosine waves. Slightly different frequencies produce the effect seen at (c)

per second, relative to the frozen ω vector (Figure 7.2(c)). Had the second sinewave been $(\omega - 2\pi)$ rad/s, that is to say 1 Hz lower than ω, then its relative rotation would have been clockwise.

The method used by the paging receiver mentioned earlier to distinguish between the equal frequency baseband tones produced when the homodyne receiver is tuned midway between the two radio frequencies is shown in Figure 7.3, a block diagram of the receiver. The incoming signal is applied to two mixers, each supplied with a local oscillator drive at a frequency of f_o, but the drive to one mixer is phase shifted by 90° relative to the other. Referring back to Figure 7.2(c), a vector rotating anticlockwise at $f_{sh}/2$ relative to f_o (where f_{sh} is the frequency shift between the two FSK tones) will come into phase with the sine component of the local oscillator, $\sin(\omega_o)$, a quarter of a cycle before coming into phase with $\cos(\omega_o)$. On the other hand, when the incoming signal is $f_s/2$ lower in frequency than f_o, then the clockwise rotation of the vector in Figure 7.2(c) indicates that it will come into phase with $\cos(\omega_o)$ a quarter of a cycle before $\sin(\omega_o)$. Now relative phases are preserved through a frequency changer or mixer, so that the audio signal in the Q channel will be in quadrature with that in the I channel. Furthermore, the audio signal in one channel will lead the other channel or vice versa, according as the incoming rf tone is above or below f_o.

The two audio paths include filters to suppress frequencies much above $f_{sh}/2$ (these must be reasonably well phase-matched, obviously) after which the signals are amplified and turned into squarewaves by comparators. As the squarewaves are in quadrature, the edges of the I channel waveform occur midway between those in the Q channel, so the D input of the flip-flop will be either positive or negative when the clock edge occurs, depending upon whether the rf tone is currently higher or lower in frequency than f_o, i.e. whether the signal represents a logic 1 or a 0. The frequencies of the two rf tones are $f_o + f_{sh}/2$ and $f_o - f_{sh}/2$ and the resultant frequencies out of the mixers are the

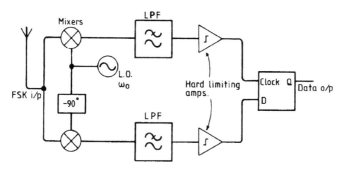

Figure 7.3 *Homodyne receiver for frequency-shift keyed transmission*

difference frequencies between these radio frequencies and the local oscillator frequency, or $(f_o + f_{sh}/2) - f_o$ and $(f_o - f_{sh}/2) - f_o$. The first of these audio tones is at a frequency of $f_{sh}/2$, while the second is at $-f_{sh}/2$ and of course by the very nature of an FSK signal, only one is present at any instant. Played through a loudspeaker they would sound indistinguishable – as indeed they are in themselves. It is only by deriving two versions of, say, $+f_{sh}/2$ using quadrature related local oscillators and comparing them that it can be distinguished from $-f_{sh}/2$. The ability of the receiver to distinguish between two audio tones of identical frequency, one positive and one negative, indicates that negative frequencies are 'for real', in the sense that a negative frequency has a demonstrable significance different from that of its positive counterpart. This can only be observed, however, if both the P and Q (in-phase and quadrature versions) are available: the signal is then said to be a 'complex' signal. A complex signal cannot be conveyed on a single wire, unlike an ordinary or 'real' signal.

FM reception

In the case of more general FM signals, including analog voice, more extensive processing of the baseband (i.e. the zero-frequency IF) signals is required. Whilst this could, in principle, be carried out in analog circuitry, it is often nowadays performed with digital signal processing (DSP) hardware. The great attraction here is that one set of digital hardware can provide any required bandwidth and any type of demodulation (rather than having separate hardware filters and detectors for AM, FM, PM, etc.) in, say, a professional or military communications or surveillance receiver (at present the arrangement would be unnecessarily expensive in a broadcast FM set). The signals must first be digitised, which at present cannot be done economically at rf with enough bits to provide sufficient resolution. A superhet front-end translates the signal, via one or more IF frequencies, to a low IF. There it can be conveniently digitised directly, or alternatively translated to zero Hz and then digitised.

There are several examples of receivers using this approach. The STC model *STR 8212* is a general coverage HF receiver with a DSP back-end which includes FM in its operating modes. In such a receiver, a non-standard IF bandwidth is easily implemented, requiring only a different filter algorithm in PROM, rather than a special design of crystal filter, with the associated design time and cost penalties. A rather similar set is available from one of the large American manufacturers of communications receivers. Another implementation of a high performance HF-band receiver with a zero-frequency final IF is described in Coy *et al.* (1990). (This did not list

FM as one of its modes, but discussion with the authors afterwards confirmed that this mode is indeed included.) At the same venue, a paper from Siemens Plessey Defence Systems Ltd (Dawson and Wagland, 1990) described their *PVS3800* range of broadband ESM receivers covering 0.5–1000 MHz. These use a DSP back-end and include an FM demodulation mode; from the brief details given it would seem likely that again a zero-frequency IF is used.

To understand the reception of conventional analog FM signals by a homodyne receiver, it is time to introduce the general expression for a narrow-hand signal centred about a frequency ω_o rad/s; this is

$$V(t) = P(t) \cos \omega_o t - Q(t) \sin \omega_o t \qquad (7.1)$$

where $P(t)$ and $Q(t)$ are called the in-phase and quadrature components. It is important to realise that equation (7.1) is only useful to describe narrowband systems, such as could pass through a bandpass filter with a bandwidth of not more than a few percent of the centre frequency; for a wideband system it would become mathematically intractable. So bear in mind that the functions of time $P(t)$ and $Q(t)$ are relatively slowly varying functions, that is to say a very large number of cycles of the carrier frequency $\omega_o/2\pi$ Hz will have elapsed by the time there has been any significant change in the values of $P(t)$ and $Q(t)$. With this proviso, equation (7.1) can, with suitable values of P and Q represent any sort of steady state signal, including FM. I am using this expression, following the development in Roberts (1977), rather than the possibly more usual approach followed by other writers (e.g. Tibbs and Johnstone, 1956) because it seems to fit in better with the explanation which follows.

Now FSK is a very specific and unrepresentative form of frequency modulation, resulting when a discrete waveform representing a digital data stream is used to modulate the frequency of a transmitter, but I introduced it first for the sole purpose of clearing up the question of the existence of negative frequencies. In the more general case, an FM signal results when a continuous waveform representing a voltage varying with time, for example speech or music, is used to modulate the frequency of a transmitter. The resultant rf spectrum is in general very complex, even for modulation with a single sinusoidal tone, unless 'm', the modulation index, is small. This is defined as the peak frequency deviation of the frequency modulated wave above or below the centre frequency (the unmodulated carrier frequency), divided by the modulating frequency. Thus, if the amplitude of a 1 kHz modulating frequency at the input of a transmitter be adjusted for a peak frequency deviation of ±2 kHz, then $m = 2$. It is fairly easy to show that, in the case of modulation by a single sinusoidal tone, the peak phase deviation from the phase of the unmodulated carrier is simply equal to ±m radians. For any modulating waveform there will be a peak frequency deviation and a corresponding peak phase deviation, but the term modulation index is only really meaningful when talking about a single sinusoidal modulating tone.

Before pursuing the niceties of the FM signal, however, I must explain the significance of $P(t)$ and $Q(t)$. If P is a constant (say unity) and Q is zero or vice versa, the result is a unit amplitude cosine or sine waveform of angular frequency ω_o (the centre frequency), the only difference being that one would be at its positive peak, the other at zero but increasing, at the instant $t = 0$, respectively. Looking at the effect of other values of the constants, if $P = Q = 0.707$ (I have written just P rather than $P(t)$ here, since $P(t)$ indicates a function of time, i.e. a variable, whereas just at the moment I am considering constants) then, as Figure 7.4 shows, the phase of the rf waveform is −45° at $t = 0$ and its amplitude (by Pythagoras' theorem) is unity. Note that the phase at $t = 0$ (or any other time, relative to an undisturbed carrier wave cosine $\omega_o t$) is given by $\tan^{-1}(Q/P)$ and the amplitude by $(P^2 + Q^2)^{1/2}$. If one insists that even when P and Q are allowed to vary, i.e. are functions of time, they shall always vary in such a way that at every instant $(P^2 + Q^2)$ is constant, then there will be a wave of constant amplitude. In this case, since the amplitude modulation index is zero, any information that the signal carries is due to variation of frequency and it can be described by the values of P and Q.

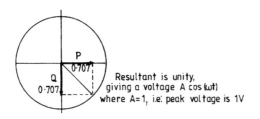

Figure 7.4 *In-phase and quadrature components. If $P^2 + Q^2$ is constant, the wave is of constant amplitude*

To start with a very simple example, suppose $P(t) = \cos \omega_d t$ and $Q(t) = \sin \omega_d t$, where $\omega_d = 2\pi$ rad/sec (say). Since $\cos^2 x + \sin^2 x = 1$ for all possible values of x (including therefore $\omega_d t$), the result is a constant amplitude signal. Further, its phase relative to ω_o is $\tan^{-1}(\tan \omega_d)$ or simply $\omega_d t$. In other words, since the phase of the signal is advancing by $\omega_d = 2\pi$ rad/s relative to ω_o, the signal is 1 Hz higher than ω_o – a (constant) deviation of +1 Hz from the centre frequency. Now if ω_d had been −2π rad/s, then the deviation would have been −1 Hz, since $\cos(-x) = \cos(x)$, whereas $\sin(-x) = -\sin(x)$. Thus the deviation is simply the rate of change of the phase of the modulated signal with respect to the unmodulated carrier. If now ω_d itself varies sinusoidally at an audio modulating frequency ω_m, then the result is a frequency modulated wave. But if, like me, you start to get confused as the algebraic symbols go on piling up, take heart; some waveforms are coming up in just a moment. However, there is one further expression to look at before we consider some waveforms, since it forms the basis of the particular form of FM demodulation to be examined.

In FM, the transmitted information is contained in the deviation of the instantaneous frequency from the unmodulated carrier – indeed, the deviation *is* the transmitted information. But the deviation is simply the rate of change of the phase angle of the signal relative to the unmodulated carrier; this phase angle is equal to $\tan^{-1}(Q(t)/P(t))$, or ϕ, say. So the instantaneous frequency of the signal is

$$\omega_i = \omega_o + d\phi/dt \qquad (7.2)$$

Now ω_o is a constant and so conveys no information: to demodulate the signal one must evaluate $d\phi/dt$, that is $d\{\tan^{-1}(Q(t)/P(t))\}/dt$. After a few lines of algebraic manipulation this turns out to be

$$d\phi/dt = \frac{P(t).dQ(t)/dt - Q(t).dP(t)/dt}{P^2(t) + Q^2(t)} \qquad (7.3)$$

Now as seen earlier, if $P^2(t) + Q^2(t)$ is constant, the result is a constant envelope wave. For an FM signal at a receiver, this condition is fulfilled (ignoring fading for the moment) so, to recover the modulation, a circuit which implements the numerator of the right-hand side of equation (7.3) is needed. Such a circuit is shown in block diagram form in Figure 7.5. Taking it in easy stages, start with Figure 7.6(a), which recaps on the basic trigonometric identity $\sin^2 \phi = (1 - \cos 2\phi)/2$, as can be seen by multiplying $\sin \phi$ by itself, point by point. Figure 7.6(b) recalls how $d(\sin a\omega t)dt = a\omega \cos a\omega t$, i.e. when you differentiate a sinewave, it suffers a 90° phase advance and the

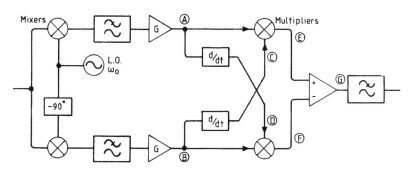

Figure 7.5 *Sine/cosine demodulator, which produces the numerator of equation (7.3) at G*

amplitude of the resultant is proportional to the frequency of the original.

In Figure 7.5, assume that P is fixed at $+2000\pi$ rad/s, and Q likewise. There is thus a fixed frequency offset of 1 kHz (2000π rad/s) above the carrier frequency ω_o. In Figure 7.5 the frequency of the incoming signal is first changed from being centred on ω_o to being centred on zero by mixing it with a local oscillator signal which is also at ω_o. The two quadrature related versions of the LO give the in-phase and quadrature baseband versions, P and Q, of the incoming signal. In the upper branch of Figure 7.5, the P or in-phase (cosine) component of the signal (now at the original modulation frequency of +1 kHz) is multiplied by a differentiated version of the Q or quadrature component. Since these are in phase with each other, the result is a waveform at twice the frequency and with a dc offset equal to half its peak-to-peak value, i.e. always positive, as in Figure 7.6(a). Figure 7.7(a) shows this and also the waveforms corresponding to the lower branch of the circuit in Figure 7.5. Here, the resultant

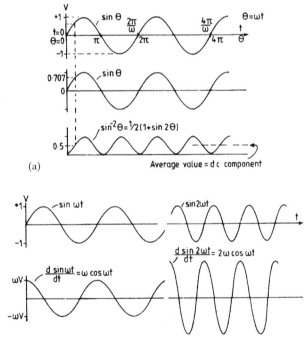

Figure 7.6 *Effects of squaring and differentiating sine waves. Squaring the wave, as in (a), doubles its frequency and produces a dc component. Differentiating, shown in (b), gives a cosine wave with an amplitude proportional to frequency*

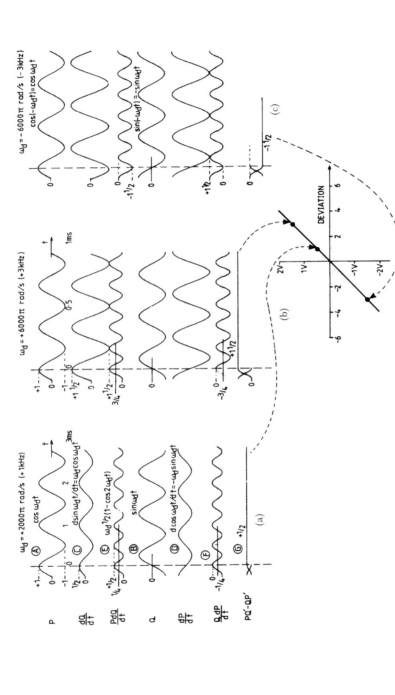

Figure 7.7 Sinewave demodulator operation with a constant frequency offset. As seen in Figure 7.5, subtracting QdP/dt from PdQ/dt gives a dc level proportional to frequency

waveform is again at twice the frequency but always negative, $d(\cos \omega_d t)dt = -\omega_d \sin \omega_d t$. Finally, subtracting $Q(t).dP(t)/dt$ from $P(t).dQ(t)/dt$, as in Figure 7.7(a), gives a pure dc level. (Note that P' is shorthand for $dP(t)/dt$.) All traces of waveforms at $2\omega_d$ wash out entirely, since when $Q(t).dP(t)/dt$ is zero $P_t.dQ(t)/dt$ is at its maximum and vice versa. Figure 7.7 also shows the results when the deviation is +3 kHz and −3 kHz, giving three points on the discriminator curve, which is a straight line passing through the origin. If ω_d, instead of being constant, varies in sympathy with the instantaneous voltage of the programme material, then the output of the circuit will simply be a recovered version of the original modulating signal as broadcast. This is illustrated for modulation by a single sinusoidal tone in Figure 7.8.

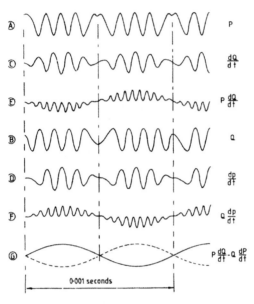

Figure 7.8 *Waveforms seen in the demodulator of Figure 7.5, with a 1 kHz FM signal of peak deviation 7 kHz*

Note that, if the *LO* frequency is not exactly equal to the carrier frequency of the received signal, then the output of the circuit will contain an offset voltage, proportional to the mistuning, but this will not in any way affect the operation of the circuit described. Indeed, in principle the offset could be equal to the peak output voltage at full modulation, so that the recovered audio would always be of one polarity, providing that the lowpass filters in Figure 7.5 had a high enough cut-off frequency to pass twice the maximum deviation frequency. The offset could be even greater; one could in theory apply equation (7.3) directly to a received broadcast FM signal at 100 MHz, using the signal direct for the $P(t)$ input and a version delayed by a quarter wavelength of coaxial cable for the $Q(t)$ input. However, with the broadcast standard peak deviation limited to ±75 kHz (mono), the peak recovered audio would amount to only 0.075% of the standing dc offset, giving a rather poor signal-to-noise ratio.

Homodyne in practice

The circuit of Figure 7.5 could be implemented entirely in analog circuitry, using double balanced mixers, lowpass filters and opamps. Differentiation is very simply performed with an opamp circuit, with none of the drift problems that beset integrators, while the multipliers could be implemented very cheaply using operational transconductance amplifiers (OTAs). An application note in the Motorola linear handbook explains how to connect the *LM13600* OTA as a four-quadrant multiplier. However, as the denominator of equation (7.3) was ignored, the output of the circuit will vary in amplitude in sympathy with the square of the strength of the incoming signal; there is no AM suppression. The amplifiers G in Figure 7.5 cannot be made into limiting amplifiers since, for the circuit to work, the base band P and Q signals need to remain sinusoidal. In principle, the amplifiers could be provided with AGC loops, but these would need to track exactly in gain: not very practical.

Alternatively, the whole of the processing following the mixer lowpass filters in Figure 7.5 can be performed by digital signal processing circuitry; the P and Q baseband signals would be popped into A-to-D converters and digitised at a suitable sample rate. This would have to be at least twice the frequency of the highest audio modulation frequency, even for narrow band FM. For wideband FM, the sampling frequency would have to be at least twice the highest frequency deviation to cope with the P and Q signals at points A and B in Figure 7.5. In practice, it would need to be higher still to allow for some mistuning of the *LO*, resulting in the positive peak deviation being greater than the negative or vice versa, and also to allow for practical rather than 'brickwall' lowpass filters following the mixers.

All the mathematical operations indicated in equation (7.3) can be performed by a digital signal processor, resulting in a digital output data stream which only needs popping into a D-to-A converter to recover the final audio. In addition to evaluating the numerator of equation (7.3) on a sample by sample basis, the DSP can also calculate $P^2(t) + Q^2(t)$ likewise. By dividing each sample by this value, the amplitude of the value of the final data samples is normalised; that is, the amplitude is now independent of variations of the incoming rf signal amplitude – AM suppression has been achieved. Naturally, this only works satisfactorily if the signals going into the A-to-D converters are large enough to provide a reasonable number of bits in the samples, otherwise excessive quantisation noise will result.

I do not know of any homodyne FM receivers working on the principles outlined in this section, in either an analog or digital

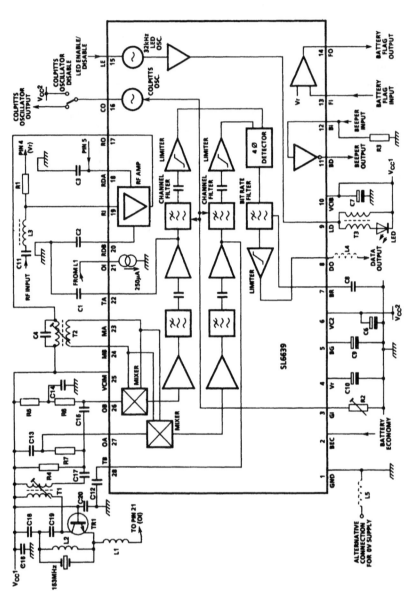

Figure 7.9 *Practical application of the SL6639 direct-conversion FSK data receiver chip from Plessey – a 153 MHz receiver for a data rate of 512 bit/s*

implementation, other than the special case of FSK paging receivers such as that described earlier. Here I am limiting the term 'homodyne' to receivers which translate the received signal directly from the incoming rf to baseband, that is to an IF of 0 Hz. In this sense, a homodyne is a heterodyne receiver, though not a 'superhet'. However, the homodyne principle as described can be and is used as the final IF stage in a double or triple superhet, the penultimate IF being translated down to the final IF of 0 Hz, and there digitised. The following DSP section provides all the usual demodulation modes, including narrow band FM, implemented as indicated using equation (7.3) in full.

References

Coy, Smith and Smith (1990) Use of DSP within a High Performance HF Band Receiver. *Proc. 5th International Conference on Radio Receivers and Associated Systems.* Cambridge, July. Conf. Publication No. 325.
Dawson and Wagland (1990) A Broadband Radio Receiver Design for ESM Applications. *Proc. 5th International Conference on Radio Receivers and Associated Systems,* Cambridge, July. Conf. Publication No. 325.
Hawker, P. (1978) Keep it simple – direct conversion HF receivers. *Proc. Conference on Radio Receivers, IERE,* July, 137.
Roberts, J.H. (1977) *Angle Modulation.* Peter Peregrinus.
Tibbs and Johnstone (1956) *Frequency Modulation Engineering,* 2nd edn. Chapman and Hall, London.
Vance, I.A.W. and Bidwell, B.A. (1982) A New Radio Pager with Monolithic Receiver. *Proc. Conf. on Communications Equipment,* IEE April.

More on long-tailed pairs

The long-tailed pair has been widely employed in circuit design ever since it first appeared. It is now widely used in double balanced mixers for rf applications.

LTPs and active double balanced mixers

The long-tailed pair (LTP) has proved a seminal influence in analog circuit design, ever since it first appeared. One of its earliest applications was at dc, in valve voltmeters (Figure 7.10(a)). Before the

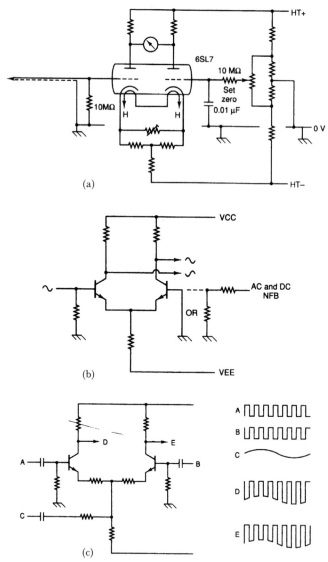

Figure 7.10 (a) A double triode providing an almost drift-free high input impedance voltmeter, greatly reducing the effect of mains voltage variations since the temperature of the two cathodes was equally affected. (b) The LTP is also useful in ac applications, e.g. this 'phase splitter', such as might be used in a push–pull amplifier. (c) The LTP can also be used as an rf modulator, but the output at each collector is 'unbalanced', i.e. contains components at both the carrier and at the baseband (modulating) frequency. If the two outputs are combined in a push–pull tank circuit, it is single balanced (containing no baseband component), but the carrier is still present, i.e. the output is AM

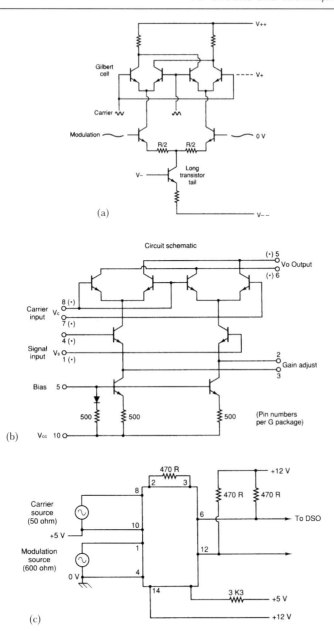

Figure 7.11 (a) The basic seven transistor tree double balanced modulator. (b) An IC version is available from many manufacturers under type numbers such as LM1496, LM1596. Note that pin numbers refer to the round G package. (c) Simple test circuit using LM1496. Note that pin numbers refer to the DIL P package

days of semiconductors, a double triode was about as close as one could get to a monolithic matched pair. In a sensitive, high input impedance valve voltmeter, designed accurately to measure dc levels, drift is a major problem. The HT circuit could be stabilised easy enough, but heater supplies were more expensive to stabilise. But with a double triode, any change in anode current for a given grid voltage in one half of the valve, due to heater voltage/cathode temperature change, would be largely cancelled by a similar change in the other half.

When transistors came on the scene, the LTP really came into its own. Early versions employed selected discrete transistors, packaged in a common heatsink to avoid temperature differences. Later, single can devices like the *2N2060* came on the scene, offering even lower drift in dc coupled circuits. The LTP is useful in a host of ac applications as well, Figure 7.10(b) showing just one. It also has many uses at rf, including that described in the previous section (see also Hickman, 1992). Figure 7.10(c) shows how the basic LTP can be used as an rf modulator. When LTPs are piled up together, things really start to get interesting.

Figure 7.11(a) shows a seven transistor 'tree', which forms the basis of many modulator/demodulator/mixer circuits. The baseband-to-rf conversion conductance is set by the value of R, the total resistance between the emitters of the lower LTP. Each of the two outputs is

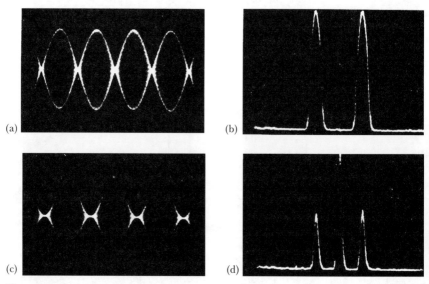

Figure 7.12 *(a) Double sideband suppressed carrier output of the MC1496. (b) Spectrum of (a), rf = 0.5 MHz, baseband modulation 1 kHz. (c) 100% modulation AM, also with rf = 0.5 MHz, baseband modulation 1 kHz. (d) Spectrum of (c)*

double balanced in its own right, containing neither carrier nor baseband components, at least if the circuit is ideally symmetrical. The circuit produces a double sideband suppressed carrier output and, used in conjunction with a suitable sideband filter, forms a simple SSB exciter. Whilst it could be built using discretes, an IC implementation provides close matching of all the components, and the provision of separate constant current transistor tails for the lower LTP enables the conversion conductance to be set with a single resistor (Figure 7.11(b)). The larger this resistor, the lower the conversion conductance, but the more linear the circuit operation becomes.

By contrast, the upper LTPs are operated without any such emitter degeneration. Their job is to switch the current smartly at their emitters to one or other collector circuit with as little delay as possible. To this end, the amplitude of the carrier is made large compared with the 100 mV or so needed to switch the upper LTPs; alternatively, a squarewave carrier can be employed. An *MC1496* was connected into the circuit of Figure 7.11(c), ready for some practical measurements, but before describing those let us take a look at some of the illustrations on the data sheet.

Figure 7.12(a) shows the DSB output of an *MC1496*, the (suppressed) carrier frequency being 500 kHz and the modulating frequency 1 kHz. The spectrum (Figure 7.12(b)) shows the carrier to be almost completely suppressed, the device providing a typical suppression of 65 dB at 0.5 MHz and 50 dB at 100 MHz carrier frequency respectively. Carrier suppression is optimised by applying a null adjustment to the bases of the lower LTP, to cancel out any standing V_{be} offset. If such an offset is deliberately introduced, then a standing carrier component will be present in the output. This permits the production of (full carrier) AM (amplitude modulation) (Figure 7.12(c)) which shows very nearly 100% modulation, the spectrum being shown in Figure 7.12(d).

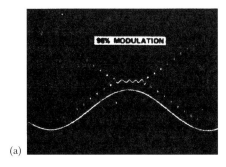
(a)
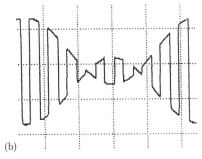
(b)

Figure 7.13 *(a) AM with modulation index of 96%. The waveform is continuous with no sudden phase changes. (b) AM with modulation index of 130%; instantaneous 180° phase flips are visible twice per cycle*

Amplitude modulation is used for broadcasting on the long, medium and short wavebands and is simply recovered from the incoming signal with a diode detector. The detector charges up a capacitor to the peak of the rf waveform, whilst a resistor in parallel with the capacitor enables the voltage to leak away again. The *RC* time constant used is long compared to the period of the rf, but short compared with that of the highest baseband frequency, enabling the detector output (hopefully, see Chapter 4 'Measuring detectors (Part 1)'; see also Hickman, 1991) to follow the peaks of the rf down into the troughs of the modulation. Due to the large disparity between the baseband and rf frequencies, the individual cycles of the latter are not visible in Figure 7.12(c), but in fact the carrier is continuous, and the waveform exhibits no phase changes. This is illustrated in Figure 7.13(a), where for clarity a much lower rf has been used; the modulation depth is 96% and the baseband waveform is shown as well for comparison. (The rf waveform here looks sinusoidal rather than square because I have cheated slightly. The waveform shown was actually produced by an *MC1495* which is a four quadrant multiplier, rather than by an *MC1496* with its cell of four switching transistors. But an active double balanced modulator like the *MC1496* would of course usually be used with a tuned tank circuit rather than with resistive loads as in Figure 7.11(c), providing a sinusoidal output.) In contrast, if the AM modulation index is allowed to exceed 100%, then there are sudden 180° phase changes apparent in the rf waveform – see Figure 7.13(b) which was produced using the circuit of Figure 7.11(c). A diode detector is only sensitive to the amplitude of the peaks, not to their phase, and so the recovered baseband audio would be a grossly distorted version of a sinewave.

Of course, broadcasters do not let the modulation index exceed 100%, although there is a tendency (especially during the mayhem on medium wave after dark) to use clippers and/or volume compressors to keep the average modulation index high. However, medium- or short-wave radio signals received from a distant transmitter are subject to frequency selective fading. The result is that the carrier can fade much more deeply than one or both sidebands, resulting in severe distortion. Furthermore, at the same time, the receiver's AGC (automatic gain control) will increase the IF gain in response to the decreased carrier level, resulting in a very loud and unpleasant noise! In principle, the distortion could be avoided by employing an active double balanced mixer to give synchronous detection, using a locally generated carrier. But unlike SSB (where a frequency error of a few cycles between the signal's suppressed carrier and the local carrier is acceptable) the local carrier would have to be at exactly the right frequency and indeed also at the right phase, i.e. it must be phase locked to the incoming carrier. In Figure 7.13(b), there is clearly still a substantial carrier component, so this would undoubtedly be possible.

An interesting case arises when the modulating frequency is the same as the rf carrier frequency F_c. The double balanced modulator or mixer then acts as a phase sensitive detector, producing the sum and difference of the modulating and carrier frequencies. As these are in this case identical, the results are $2F_c$ and 0 Hz, i.e. the second harmonic and a dc term. The value of the latter term can be anything from a peak positive value, through zero, to the same peak value but negative, depending upon the relative phasing of the inputs at the carrier and modulation ports of the mixer. Figure 7.14(a) shows a circuit where the two inputs are in phase, which results in a peak positive output at one output port of the mixer and peak negative at the other. If, moreover, a large signal is used so as to overdrive the modulation port, the second harmonic output becomes small, one output remaining 'stuck high' and the other 'stuck low'.

The exceptions to this are the two instants each cycle where the input waveform passes through zero, at which points all transistors must be conducting equally. This results in narrow spikes (Figure 7.14(b)), which are rich in harmonics (the slightly rounded shape of alternate half cycles is due to the single ended drive to the modulation port). This circuit can thus be used as a frequency multiplier: for example, using a 5 MHz input from a standard frequency source, the outputs could be combined in a tuned push–pull tank circuit so as to extract any desired harmonic of 10 MHz. With further overdriving of the modulation port, or by using a clipped sinewave drive, the spikes become very narrow, enabling very high order harmonic generation to be simply achieved. If the input rf is

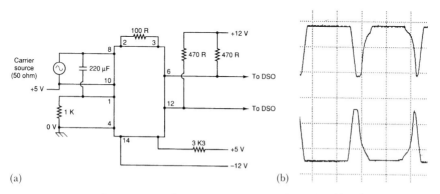

Figure 7.14 (a) Circuit used for harmonic generation. (b) Waveforms at the outputs of the circuit shown in (a).

frequency modulated, the deviation will be increased pro rata to the order of the selected harmonic. Thus phase modulating the original oscillator with baseband audio which has been subjected to a 6 dB/octave top cut, and then selecting a harmonic at VHF, would provide a comparatively simple NBFM generator, all done with LTPs.

Acknowledgements

Figures 7.11(b), 7.12 and 7.13(a) are reproduced by courtesy of Motorola.

References

Hickman, I. (1991) Measuring detectors. *Electronics World + Wireless World*, November, 976.
Hickman, I. (1992) Oscillator tails off lamely? *Electronics World + Wireless World*, February, 168.

'No-licence' transmitters

There is considerable interest in low power radio telemetry and related applications following recent UK deregulation. The DTI no longer requires licensing for certain types of low power radio link provided that the equipment meets an approved specification and is used in accordance with the regulations. Wireless data links have never been easier.

Low power radio links

World War II saw an explosion in the applications of radio/radar, and since then further applications have proliferated. For the man in the street, the most notable milestones were probably television, FM broadcasting and colour television, in that order, but many other users have come to regard wireless communication as indispensable to their everyday business. Examples are PMR (private mobile radio, used by taxi and delivery firms, despatchers for service industries of all sorts, etc.) and more recently car telephones, whilst further services such as GSM, DECT, Cordless II, Phonepoint, etc. are either waiting in the wings or happening now.

One area which has seen considerable growth is low power telemetry and related applications, partly due to a measure of deregulation in the 1980s. The Low Power Devices Information Sheet BR114, from the Radiocommunications Division of the DTI and dated May 1989, listed in Annex 1 a number of types of low power devices for telemetry, etc. that were exempted from licensing by the end user, though, of course the manufacturer was still required to obtain type approval for the device to the appropriate MPT specification before offering it for sale. Telemetry is defined as the use of telecommunication for automatically indicating or recording measurements at a distance from the measuring equipment, whilst the related applications include: Telecommand, the use of telecommunication for the transmission of signals to initiate, modify or terminate functions of equipment at a distance; Teleapproach, the use of telecommunication for the purpose of gaining information as to the presence of any moving object; Radio Alarm, an alarm system which uses radio signals to generate or indicate an alarm condition, or to set or unset the system; Radio Microphone, a microphone which uses a radio link to convey speech or music to a remote receiver; and sundry other uses including induction systems for the hard of hearing in cinemas and other public places, metal detectors, model control, access and antitheft devices and passive transponder systems.

BR114 also specified (in Annex 2) a further list of low power devices that were not exempt from licensing. This was coupled with a note to the effect that the Department intended to exempt some of these at some future date, and in the meantime would issue licences free of charge, and even supply manufacturers of type approved equipment with blank licences for them to issue as required! BR114 has been superseded by RA114, dated July 1991, from the reorganised Radiocommunications Agency, and alarms in general (those covered by MPT Specifications 1265, 1344, 1360, 1361 and 1374) have been transferred to Annex 1, the exempt category. This is reproduced here as Table 7.1 and it can be seen that the allocations span the spectrum from VLF (e.g. 0–185 kHz, induction communications systems) through HF, VHF and UHF to centimetric wavelengths (e.g. various allocations between 2.445 and 33.4 GHz, low power microwave devices). The Annex 2 items, not exempt from licensing, are shown in Table 7.2. Some of the frequency allocations are the same as, or adjacent to Annex 1 allocations, the difference being that in general the licensed devices are permitted a higher ERP.

A few years ago, I was asked by an old friend whose company manufactures personal alarms for the elderly and infirm, if I could design a short range radio pendant, so that the wearer could summon help from anywhere in the house or garden. I replied that I could, but

Table 7.1 Exempt devices

Use	Frequency	Maximum ERP	Specification
Induction communications systems	0–185 kHz and 240–315 kHz	See specification, transmitter output is 10 watts maximum	MPT 1337
Metal detectors	0–148.5 kHz	See SI no. 1848/1940	N/A+
Access and antitheft devices and passive transponder systems	2–32 MHz	See specification	MPT 1339
Telemetry, telecommand and alarms			
General telemetry and telecommand	26/27 MHz	1 mW	MPT 1346
Short range alarms for the elderly and infirm	27/34 MHz	0.5 mW	MPT 1338
General telemetry and telecommand (narrow band)	173.2 to 173.35 MHz	1 mW*	MPT 1328
General telemetry and telecommand (wide band)	173.2 to 173.35 MHz	1 mW*	MPT 1330
Short range fixed or in-building alarms	173.225 MHz	1 mW	MPT 1344
General telemetry, telecommand and alarms	417.90 to 418.10 MHz	250 µW	MPT 1340
Industrial/commercial telemetry and telecommand	458.5 to 458.8 MHz	500 mW	MPT 1329
Alarms			
Car theft paging alarm	47.4 MHz	100 mW	MPT 1374
Radio alarms (marine alarms) for ships	161.275 MHz	10 mW	MPT 1265
Mobile alarms	173.1875 MHz	10 mW	MPT 1360
Fixed alarms – above 1 mW and up to 10 mW	173.225 MHz	10 mW	MPT 1344
Fixed alarms	458.8250 MHz	100 mW	MPT 1361
Transportable and mobile alarms	458.8375 MHz	100 mW	MPT 1361
Car theft paging alarms	458.9000 MHz	100 mW	MPT 1361
Model control			
General models	26.96 to 27.28 MHz	100 nW	N/A+
Air models	34.995 to 35.225 MHz	100 mW	N/A+
Surface models	40.665 to 40.955 MHz	100 mW	N/A+
Other			
General purpose low power devices	49.82 to 49.98 MHz	10 mW	MPT 1336

Radio microphones and radio hearing aids	173.35 to 175.02 MHz	5 mW (narrow band) 2 mW (wide band)		MPT 1345
Low power microwave devices	2.445–2455 GHz	100 mW		MPT 1349
	10.577–10.597 GHz	1 W		
	10.675–10.699 GHz	1 W		
	24.150–24.250 GHz	2 W		
	24.250–24.350 GHz	2 W		
	31.80–33.40 GHz	5 W		

Table 7.2 *Non-exempt devices*

Use	Frequency	Maximum ERP	Specification	Application form
Audio frequency induction loop deaf aid systems* (higher power non-carrier systems)	0.16 kHz	See specification – transmitter output is above 10 watts	MPT 1370 (in draft)	RA77
Telemetry and telecommand				
Medical and biological telemetry	300 kHz to 30 MHz	See specification	W6802/MPT 1356 (in draft)	RA77
Telemetry systems for data buoys	35 MHz	250 mW	MPT 1264	RA61
General telemetry and telecommand (narrowband) – above 1 mW and up to 10 mW	173.20 to 173.35 MHz	10 mW	MPT 1328	RA77
General telemetry and telecommand (wideband) – above 1 mW and up to 10 mW	173.20 to 173.35 MHz	10 mW	MPT 1330	RA77
Medical and biological telemetry (narrowband and wideband)	173.7 to 174.0 MHz	10 mW	MPT 1309/ MPT 1312	RA77
Medical and biological telemetry	458.9625 to 459.1000 MHz	500 mW	MPT 1363 (in draft)	RA77
Teleapproach				
Teleapproach (perimeter intruder detection systems)	40 MHz or 49 MHz	See specification	MPT 1364 (in draft)	RA77
Teleapproach antitheft devices	888/889 MHz (and 0 to 180 kHZ)	See specification	MPT 1353 (MPT 1337)	RA50

© Crown Copyright 1991, Radiocommunications Agency

that we would still be faced with the need to obtain type approval, and therefore pointed him in the direction of a manufacturer of existing type approved transmitter modules and matching receivers (Radiometrix Ltd, see Ref. 2). Such modules are produced by many manufacturers and a number of these have banded together with designers and users in a trade association, the LPRA (Low Power Radio Association, see Ref. 3, from which a membership directory is available) with the aim of promoting high standards in the design and use of such modules and systems. Whereas many of the modules produced by manufacturers for this market area are designed with a specific end use in mind – such as security systems – the modules which I recommended to the manufacturer, and which are described in further detail below, are totally uncommitted. This means that the purposes to which they can be put are limited only by the user's ingenuity, the mode of operation being determined by the nature of the peripheral circuitry with which he surrounds them. The modules in question operate in the band 417.90–418.10 MHz and are type approved to MPT 1340: *General Telemetry, Telecommand and Alarms*, which specifies a maximum ERP of –6 dBm, i.e. 250 µW.

Given a pair of such modules to experiment with, I was naturally keen to see what I could find out about them with the limited amount of equipment then available in my home laboratory. Consequently, the 'TXM-UHF' transmitter module was, for speed and convenience, mounted on an odd piece of strip-board as a means of connecting power, modulation, etc. (see Figure 7.15(b)), Figure 7.15(a) shows the block diagram of the transmitter module. Of course, anyone who contemplates constructing a UHF circuit on strip-board needs his head examining, but in this case all the 'hot' circuitry was on the module, with only dc and low frequency inputs supplied via the strip-board. The rf output was connected to a couple of inches of 50 Ω coax, the other end of which was terminated in a BNC plug. This was connected to the channel 1 input of a Tektronix 475a oscilloscope via a 20 dB attenuator. The transmitter was powered from a 12 V dc, its maximum rated input voltage, and the trace on the scope photographed (Figure 7.15(c)). The trace clearly shows more than four but less than four and a quarter complete cycles across the screen, so the most one could say about the frequency is that, if you believe the 'scope's timebase accuracy implicitly, the frequency could well be 418 MHz. The rated output (max.) of the unit at 12 V is –3 dBm ERP, which would correspond to 159 mV rms into a 50 Ω load, if that were what the module were designed to drive (the recommended antenna is a quarter wave whip, which, above a ground plane, would present an on tune impedance of 35 Ω).

The 50 mV pp of Figure 7.15(c) corresponds to 177 mV rms at the input of the 20 dB pad, which seems unlikely at best, not least

RF circuits and techniques 293

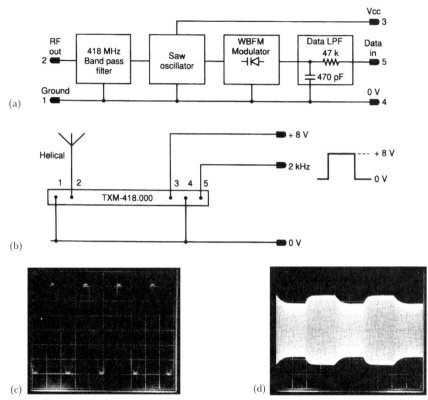

Figure 7.15 *(a) Block diagram of the TX-UHF transmitter module. (b) Connections in recommended test circuit. (c) The output of the transmitter module following 20 dB of attenuation, 10 mV/div. vertical, 1 ns/div. horizontal, 12 V supply, no modulation. (d) As (c) except 9 V supply, 100 µs/div. horizontal, 2.4 kHz squarewave modulation applied*

because the rated bandwidth of the 'scope is only 250 MHz. However, there are other factors to take into account: principally the input impedance of the 'scope. At low frequencies this can be represented as a lumped impedance of 1 MΩ in parallel with 20 pF, and 20 pF at 418 MHz has an impedance of –19 Ω. However, the input circuitry is anything but lumped and at 418 MHz its input impedance could be anything – quite possibly approaching an open circuit. In this case, the output of an unterminated 20 dB pad would be not 15.9 mV rms but 31.8 mV rms or 90 mV pp. An indicated answer of 50 mV pp is therefore not unreasonable, since the frequency response roll-off of a 'scope designed faithfully to reproduce fast step waveforms has perforce to be gradual – a rapid roll-off would inevitably be

294 Analog circuits cookbook

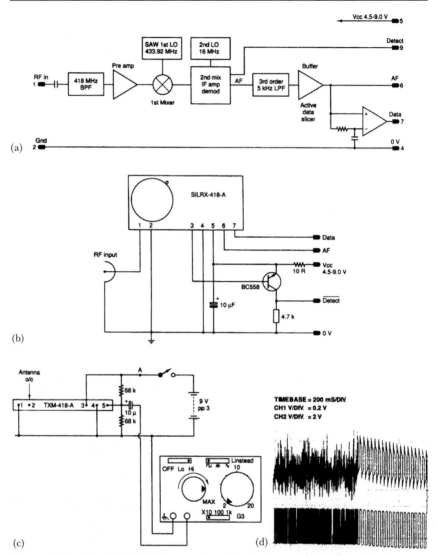

Figure 7.16 *The SILRX-418-A receiver block diagram. (b) Basic receiver test circuit (see text). (c) The transmitter circuit used with (b). (d) The receiver audio (upper trace) and data (lower trace) outputs before and after transmitter switch-on*

accompanied by a group delay distortion, resulting in ringing on fast pulse waveforms.

Nevertheless, even though the trace in Figure 7.15(c) looks like a very nice sinewave, it must be said that at 418 MHz, the 'scope would show even a squarewave as pretty well sinusoidal: although a useful amount of information can be gathered from an oscilloscope used way beyond its ratings, clearly there are limits. Figure 7.15(d) shows the rf output again, but this time at 100 μs/div. 9 V supply, with a 0 to 8 V CMOS 2.4 kHz squarewave modulation input which was also used to trigger the 'scope. The modulation is FM, produced by means of varactor diode, but it can be seen that there is noticeble incidental AM, doubtless due to the higher Q of the varactor at the higher reverse bias level of 8 V. A 2.4 kHz squarewave modulation corresponds to 'revs' (reversals, i.e. a continuous 101010... pattern) at 4.8 kbit/s, the incidental AM making the baseband lowpass filtering of the modulation clearly visible. This is incorporated in order to limit the transmitter's OBW (occupied bandwidth) by suppressing the higher order FM sidebands.

Once all the measurements on the TX which were possible with a 'scope had been taken, the TX output was connected to a divide-by-100 prescaler (Hickman, 1992), the output of which was connected to a Philips *PM2521* Automatic Multimeter, set to the frequency counter model. With a TX dc supply of 9 V and the modulation input strapped to 0 V, the frequency was 417.96 MHz and strapped to +9 V was 418.01 MHz. This was well within the maker's initial frequency accuracy specification, though since the interval since the *PM2521* was last calibrated is considerable, the absolute accuracy of the measurement cannot be relied upon completely. The incremental accuracy is reliable enough though, and clearly the transmitter's deviation is ±25 kHz. It was time now to look at the matching receiver.

The *SILRX-418-A* receiver is a double superhet design, the block diagram being used as in Figure 7.16(a). The receiver was mounted on another scrap of strip-board ready for testing in conjunction with the transmitter. The circuit was as in Figure 7.16(b), except that a *BC214* was used instead of the *BC558* and its 4.7 kΩ collector load was replaced by an 820 Ω resistor in series with an LED. The transmitter was as in Figure 7.16(c), with 20 Hz squarewave modulation from a battery powered audio function generator. As the output of the latter was centred about ground, a blocking capacitor and bias chain was employed to keep the modulation swing at pin 5 of the transmitter module within the range 0 to 8 V. In view of the minimal separation – the receiver was within a metre of the transmitter on the lab bench – no antennas were used. Figure 7.16(d) shows the receiver audio output (upper trace) and the data output (lower trace). The transmitter was switched on half way through the trace, the audio and data outputs up to that point being just noise and clipped noise respectively. Following switch-on, the audio output is a 20 Hz squarewave exhibiting considerable sag, due to partial differentation by the inadequate modulation coupling time constant. There is also an initial transient dc level shift, as the 10 μF blocking capacitor charges up. The data output is a cleanly sliced version of the audio, with the initial transient also suppressed.

In applications such as telecommand, whilst the transmitter only needs to be powered up when it is desired to send a command, the receiver must usually be ready to receive it at any time (the rare exceptions being systems where commands need only to be sent at predetermined times). However, if the receiver is battery operated, it is undesirable to have it drawing current all the time, so it is often arranged to come on briefly to look for a signal, with a duty cycle of about 1% on-time in the absence of signals. If the presence of a signal is detected (a matter of two or three milliseconds from switch-on), the DETECT signal can be used to extend the on-time to receive a command. The data settling time (time from valid carrier detect to stable data output) is another important operational parameter. These times are too short to determine from Figure 7.16(d), so the modulation frequency was increased to 200 Hz, the top of the modulation bias chain was moved from point A to the positive pole of the battery and a lever-arm skeleton microswitch, used as a push button, substituted for the on/off switch. After a few tries, the shot shown in Figure 7.17 was captured (no film was wasted; the waveforms were captured on a Thurlby-Thandar *DSA524* sampling adaptor first, and then photographed from the screen of the 'scope, which was used in this instance simply as a monitor display). The upper trace (5 V/div.) shows the switch-on of the transmitter's +9 V supply, exhibiting over two milliseconds of switch bounce, whilst the lower trace

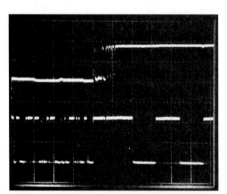

Figure 7.17 *Receiver data settling time test, upper trace (5 V/div.) TX supply switch-on; lower trace (2 V/div.) RX data output (TX modulated at 200 Hz), 2 ms/div. horizontal*

shows the receiver's recovered data output. The first negative-going edge about two milliseconds after the end of the TX switch bounce looks a bit suspect, but thereafter things are fine so clearly the data settling time is well within the maker's figure of 10 ms (although that is quoted with a 5 V receiver dc supply against the 9 V used here).

Figure 7.16(d) shows the receiver audio output reflecting the detailed shape of the transmitter modulation, indicating that the modulation and demodulation processes are fairly linear. To see just how linear, the TX was modulated with a 7.5 V pp sinewave (Figure 7.18(a), lower trace), and the RX audio captured (upper trace) for comparison. It looks just a little bit 'secondish', the positive peaks too rounded and the negative too peaky, but clearly the link would be capable of transmitting analog data. Figure 7.18(b) shows the recovered audio (upper trace) when the modulating sinewave was reduced to 3 V pp and clearly the distortion is very low indeed – at the sacrifice of some 8 dB path loss capacity. On the other hand, with the reduced deviation and consequently reduced input to the data recovery circuit, the slicer is finding it rather hard work (data output, lower trace). However, with a reduction in path loss capability of just 6 dB relative to binary modulation, it would be perfectly possible to operate the link with four levels rather than just the two of ±25 kHz deviation, preferably with some linearisation in the modulator to give equally spaced levels at the receiver. Two bits per symbol could therefore be transmitted, doubling the maximum bit rate throughput.

With the proliferation of transmission systems requiring no licence, it might seem that mutual interference would make them unusable, the more so since many of the bands are too narrow to admit of channelling, so that all devices work on nominally the same carrier frequency. In practice, a number of factors present this from being a real problem, at least for the present.

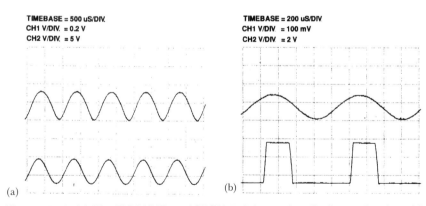

Figure 7.18 *(a) The* TXM-UHF *and* SILRX-418-A *can handle linear signals, with some distortion; 1 kHz full amplitude sinewave modulation applied to TX module (lower trace) and as recovered by the receiver (upper trace). (b) At a reduced modulation level (upper trace) the distortion of the received signal is negligible, although the effect of a reduced level into the data slicer is evident (lower trace)*

Firstly, the ERP is purposely limited to a fairly low level. Thus whilst the modules featured here might give a range of several kilometres under exceptional circumstances – with elevated antennas on a large flat plain without trees or other obstructions – the manufacturer quotes a reliable maximum range over open ground (with antennas mounted at a height of only 1.5 m) of 200 m, whilst excessively obstructed paths (with buildings etc.) and/or antennas less efficient than $1/4$ wave whips may in extreme cases reduce the reliable operating range down to some 30 m.

Secondly, the devices are designed (for the most part) for intermittent operation, e.g. in telecommand applications. Even a telemetry application will not normally broadcast continuously, but will send batches of readings at predefined intervals or on telecommand.

Thirdly, even though a receiver may pick up a transmission not intended for it, these devices are commonly operated with an address code as a header to each transmission, and a receiver can thus ignore a transmission not labelled with its own particular address code.

The first two points reduce the possibility of a wanted transmission being jammed by an unwanted, and the third minimises the possibility of inappropriate response to the reception of an unwanted signal.

Whilst NRZ (non-return to zero) data, typified by the reversals illustrated in Figures 7.16 and 7.17, could be used, a popular and commonly employed mode of signalling used with low power radio modules uses both 0 and 1 logic levels for each data bit, making the code (like Manchester code) self-clocking. A typical example is the Motorola range of CMOS devices *MC14026* (Encoder) and *MC14027/028* (Decoders). These 16 pin devices have nine pins dedicated to setting address and/or data bits. Each pin can be connected to ground (low) to V_{dd} (high) or can be left open-circuit. Thus data is trinary, permitting in principle the transmission of $3^9 = 19\,683$ different codes. The *MC14027* interprets the first five trinary bits as address giving $3^5 = 243$ different addresses, the remaining four bits being interpreted as data. For the four data bits, an open is interpreted as a logic 1, so only $2^4 = 16$ different messages are available. The *MC14028* interprets all nine pins as addresses, but with the same limitation on address pin 9 as the data pins on the *MC14027*: consequently $2 \times 3^8 = 13\,122$ different addresses are available, but only a single data bit (received or not received), indicated by the VT (valid transmission) flag. With either receiver, two consecutive valid addresses followed by identical data must be received before the new data is latched and the VT flag set.

DIL switches with a choice of three ways per pole are rather rare and so another popular scheme, typified by the 18 pin DIL plastic devices by Holtek, type number *HT12E* (12 bit encoder) and *HT12D* (8 bit address and 4 bit data decoder) uses address/data pins with binary selection. Each bit of the modulating signal consists of a low level followed by a high level: in the case of a 0 the low level persists for two-thirds of the data bit period, switching to the high level for the final one-third, whilst for a 1 the level is low for the first third of a bit period and high for the last two-thirds. (At least, that is how I have described it here, in order to be consistent with what follows – see Figure 7.19. The data sheet actually defines it the other way round. This comes about because the address and data pins have internal pull-ups, so that an 'on' (1) setting on the DIL switch pulls the corresponding pin to ground (0), and vice versa.) The twelve transmitted bits produced by the *HT12E* are preceded by a 0 as a start bit and followed by a logic low level lasting another 12 bit periods. The sequence is initiated by a low level on the TE pin and repeated four times: at the receiver, the address must be received correctly on each occasion and must be followed by identical data bits each time before the VT flag goes high. If pin 14 of the CMOS device is held low, the sequence of four blocks of 12 bits is transmitted

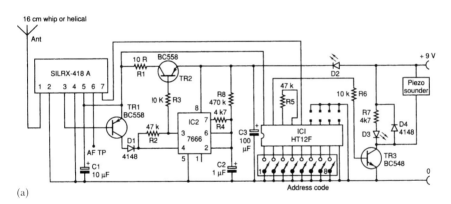

(a)

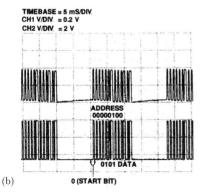

(b)

Figure 7.19 (a) SILRX-418-A receiver audio output (upper trace) and data output (lower trace) when receiving repeated 12 bit sequences of 0000 0100 0101. (b) SILRZ-418-A pager application circuit treating all twelve bits as addresses, giving 4096 different possibilities

repeatedly. At the receiver, the falling edge of the start bit indicates the start of the first address bit, the *HT12D* providing 256 different addresses and four recovered data bits. The resultant data stream at the receiver is illustrated in Figure 7.19(b), with the audio output on the top trace and the data on the lower. Following the start pulse, it can be seen that the address is set to 0000 0100 and the data to 0101. The signal was received on the *SILRX-418-A* as in Figure 7.16, but the transmitter used was one from the Radiometrix evaluation kit; this transmitter includes, in addition to a *TXM-UHF* transmitter module, an *HT12E* encoder IC, 8 way and 4 way DIL address and data switches, etc. The basic *SILRX-418-A* receiver module was not actually doing anything with the recovered data, but a simple 1-bit pager application circuit, indicating when a valid address is received, is shown in Figure 7.19(a). It includes a 1% duty cycle (4 ms on, 400 ms off) battery saving feature, the on period automatically being extended for the duration whilst a signal is present, although the LED D3 will not light nor the sounder sound unless the received address matches the address set up on the receiver. The 12 bits output from the *HT12E* are uncommitted and the *HT12F* decoder used in the circuit in Figure 7.19(a) treats all twelve as addresses, giving 4096 different possible addresses. In contrast, the receiver unit in the evaluation kit uses a slightly larger and more sophisticated RX module which works in conjunction with an *HT12D* decoder. It thus recovers 4 data bits, as well as indicating various status conditions such as signal detect, jamming detect, valid code detect and tamper alarm. With only 256 different possible addresses, it might be thought that an address bit error, due to a pulse of interference or a momentary fade (if either the TX or RX is moving), might cause a receiver to respond to a signal not intended for it. But the requirement to receive four consecutive identical addresses makes the odds against this 256^4 to 1, i.e. not very likely. Even the odds against latching wrong data are 16^4 or 1 in 65 536. In fact, the system is much more foolproof than this, since each of the four address/data blocks must be preceded by a low level lasting 12 bit periods, which will only be so if the received signal strength is adequate to provide quieting at the receiver.

The foregoing exhausted the tests that could be carried out in the home laboratory, but one or two crucial points of interest, such as the transmitter's OBW, remained. These could only be settled with the aid of a spectrum analyser, so arrangements were made to carry out further tests at the premises of the manufacturer of the modules. The close-in spectrum of the transmitter when transmitting 30 Hz squarewave modulation at ±25 kHz deviation is shown in Figure 7.20(a), indicating an OBW of less than 120 kHz at the −50 dB level.

RF circuits and techniques 301

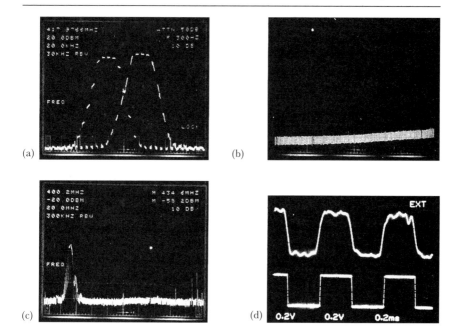

Figure 7.20 *(a) Close-in spectrum of the TXM-UHF transmitter module modulated with a 30 Hz squarewave with ±25 kHz deviation, showing an OBW of 120 kHz at the –50 dB level. (b) 0–1800 MHz spectrum, showing 2nd and 3rd harmonics more than 60 dB down and 4th harmonic over 50 dB down relative to the 418 MHz output (which is indicated by the marker just below top-of-screen reference level). (c) Spectrum of 433 MHz 1st LO radiation of the matching SILRX-418-A receiver module (at marker) and of a super-regenerative receiver operating at about 330 MHz (where there is no UK frequency allocation). Note that even for equal signal levels, the total interference power radiated from the super-regenerative receiver would be much greater since it produces lines spread over a considerable bandwidth. In fact, the antenna used on the spectrum analyser is about 6 dB down at 330 MHz relative to 433 MHz, so the single spectral line of LO radiation from the superhet receiver is at a lower level than the peak of the broad band of radiation from the super-regenerative receiver. (Low level signals at the right-hand side are Band IV TV signals.) (d) The output of the SILRX-418-A receiver with a 1 kHz squarewave modulated input signal of ±25 kHz deviation at a level of –113 dBm, i.e. 0.5 µV. Upper trace: audio output, 0.2 V/div., 500 µs/div.; lower trace: recovered data output, 2 V/div. indicating the remarkable sensitivity of the double superhet design. $P_{transmit}/P_{receive} = 107$ dB nom. or 5×10^{10} $P_t/P_r = (2.44 \pi d/\lambda)^2$, giving a theoretical path loss capability between isotropic antennas in free space of 21 km*

Figure 7.20(b) shows the far-out spectrum, with all harmonics greater than 50 dB down, thanks to the transmitter's effective output bandpass filter. Figure 7.20(c) shows the receiver's first *LO* (local oscillator) radiation as received on a 418 MHz whip, and for comparison, the radiation from a super-regenerative receiver operating at about 330 MHz (where there is no UK allocation for such devices) is also shown. It is the receiver unit of a remote radio doorbell which is widely offered for sale in the UK by postal mail order.

Super-regenerative receivers performed useful service in the Second World War but thankfully faded from the scene afterwards. They are unpleasant devices, transmitting a broad band of interference centred on their receive frequency. Nevertheless they are reappearing in short-range applications such as garage door openers, on account of their very low cost, due to the minimal circuitry required. They are disparagingly known in the trade as 'hedgehogs', an apt description once one has seen the spectrum. Needless to say, devices offered for sale by members of the LPRA, such as those featured in this article, will be well engineered and legal.

References

1. Hickman, I. (1992) A low cost 1.2 GHz prescaler. *Practical Wireless*, August, 18–23.
2. Radiometrix Ltd, Hartcran House, Gibbs Couch, Carpenders Park, Watford, Herts WD1 5EZ, UK. Tel: 0181 428 1220; fax: 0181 428 1221.
3. The Low Power Radio Association, Brearly Hall, Luddendon Foot, Halifax, HX2 6HS, UK. Tel: 0142 288 6463; fax: 0142 288 6950.

> Noise comes in all shapes and sizes (and colours!). This article looks at some of the many varieties, and their fascinating properties.

Noise

Noise is all around us. The acoustic variety is often intrusive, but the electrical sort mostly does not worry the man in the street. Except when unsuppressed cars pass too near his TV aerial, ruining the picture, or a noisy line prevents him hearing the person at the other end of the phone. But for the electronics engineer, it is a different matter. Obviously, the communications engineer is concerned, be his work in line or wireless communications. But light-current engineers in all fields are affected, since their work inherently involves the transport and processing of information by electrical means, unlike their heavy-current peers in power engineering – where the generation and distribution of electrical energy *per se* is an end in itself.

Noise – the basics

Noise comes in many guises – thermal, gaussian, baseband, broadband, narrowband, stationary, white, pink, impulsive, blue, red, non-stationary and a few others as well.

Thermal noise (also called Johnson noise or resistor noise) is inherently present in all systems operating at a temperature in excess of absolute zero (0K or –273°C). In a conductor, the electrons are in continuous random motion, in equilibrium with the molecules of the conducting material.

The mean square velocity of the electrons is proportional to the absolute temperature. As each electron carries a negative charge, each electron trajectory between collisions with molecules constitutes a brief pulse of current. As could be expected, the net result of all this activity is observable as a randomly varying voltage across the terminals of the conductor. Obviously the mean value (dc component) of this voltage is zero, otherwise electrons would be piling up at one end of the conductor, but there is an ac component, described by the Equipartition Law of Boltzmann and Maxwell.

This states that for a thermal noise source, the available power $p_n(f)$ in a 1 Hz bandwidth is given by

$$p_n(f) = kT \text{ (watts/Hz)} \tag{7.4}$$

where k = Boltzmann's constant = 1.3803E-23 (joule/K) and T is the absolute temperature of the noise source in degrees Kelvin. At room temperature (290K or 17°C) this turns out to be

$$p_n(f) = 4.00\text{E-}21 \text{ (watts/Hz)} = -204 \text{ dBW/Hz}^2 = -174 \text{ dBm/Hz}^2 \tag{7.5}$$

In $p_n(f)$, the (f) indicates that the noise power per unit bandwidth is, in general, a function of frequency. In the case of thermal noise, the power per unit bandwidth is in fact constant, so thermal noise is described (by analogy with white light, which contains components at all frequencies or colours) as 'white'.

At room temperature the value of $p_n(f)$ quoted at (7.5) is found to hold up to the highest microwave frequencies at which it has been possible to measure it. But if the bandwidth were truly infinite, the equipartition theory would predict that the power available from a thermal source would be infinite. The solution to this paradox is provided by the application of quantum mechanics, which theory requires the kT of the equipartition theory to be replaced by $hf/(\exp\{hf/kT\} - 1)$, where h = Plank's constant = 6.623E-34 (joule . seconds). This results in a modified expression for $p_n(f)$

$$p_n f = \frac{hf}{\exp\left(\dfrac{hf}{kT}\right) - 1} \quad \text{(watts/Hz)} \tag{7.6}$$

Expression (7.6) results in thermal noise actually tailing off at very high frequencies, and this is illustrated in Figure 7.21. This shows that the spectral density of thermal 'white' noise from a source at room temperature has fallen to about 90% of the low frequency value by about 1250 GHz. But for a low temperature amplifier such as a maser operating at one degree above absolute zero, the thermal noise is already 10% down at just 5 GHz.

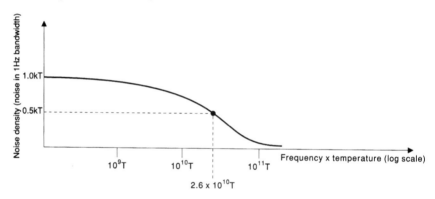

Figure 7.21 The level of thermal 'white' noise falls off above a certain frequency which depends upon the temperature

Thermal noise model

Figure 7.22 shows how a resistive noise source may be modelled. Maximum noise power is delivered to R_1 when its value equals R, but there is no *net* transfer of power. Because R_1 in turn delivers an equal amount of noise power back to R. Note that in Figure 7.22, v_n is that component of the noise appearing across R_1 *due to the noise voltage e_n of the source R only*. As measured, v_n will be larger than this, due to the component of noise across R due to the thermal noise of R_1. There is no correlation between this component and the component across R_1 due to R. Consequently, the voltage v_n' actually measured across R (or R_1), will be the *rms sum* of the two components. So in general

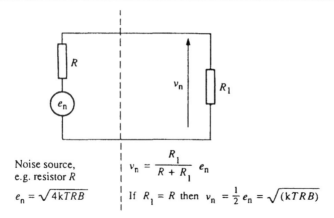

Figure 7.22 The thermal noise of a resistor R can be modelled as a noise source e_n in series with it

$$v_n' = \sqrt{\left(\frac{R_1}{R_1+R}e_{n1}\right)^2 + \left(\frac{R}{R+R_1}e_{n2}\right)^2}$$

and so if $R_1 = R$, then $v_n' = 1.414 v_n$

R may be for instance the source resistance of an antenna, so that a wanted signal e_s appears in series with e_n. The ideal signal-to-noise ratio available is thus e_s/e_n. R_1 may be the input resistance of an amplifier. In the matched case where $R_1 = R$, the amplifier sees an input signal $e_{input} = e_s/2$. But the effective source resistance is now R in parallel with R_1, or effectively $R/2$ in the matched case. So the matched-input amplifier sees not $e_n/2$ at its input but

$$\sqrt{kTB\frac{R}{2}} = \frac{e_n}{\sqrt{2}}$$

Thus the matched case incurs a 3 dB noise figure, even if the amplifier itself is noiseless.

If the amplifier has a high input impedance, so that R_1 is much greater than R, the theoretical stage noise factor $(R_1 + R)/R_1$ can approach unity (for a noise-free amplifier). The (relatively) low resistance of the source effectively shorts out the noise of the amplifier's high input resistance.

Characteristics of random noise

A source of noise may, or may not, be white like the thermal noise considered above. But most sources of noise, including thermal noise,

Figure 7.23 *A short sample of broadband noise*

exhibit the same shape of noise voltage probability density, NVPD. Figure 7.23 shows a sample of the variation of baseband noise over a period of time. The greater part of the time, the voltage is not greatly different from the mean value of zero, but peaks of either polarity occur, the larger the value of the peak the less frequently it is observed. This distribution is described as Gaussian, and the probability of the occurrence of any particular instantaneous value of voltage e_n is given by

$$e_n = \left(\sigma_n \sqrt{2\pi}\right)^{-1} \exp\left(\frac{-e_n^2}{2\sigma_n^2}\right) \qquad (7.7)$$

This expression is plotted as the Gaussian or Normal distribution in Figure 7.24, which indicates that however large a peak voltage you care to specify, if you wait long enough it will eventually occur. (However, the exponential function is a very powerful one, so that the likelihood of the occurrence of a peak of, say, twice the amplitude of the largest shown is exceedingly remote.)

The value σ_n in expression (7.7) is the standard deviation of the voltage from the mean. In practice, the mean is usually zero – as in the case of thermal noise. The noise may incidentally be riding a dc level, as at the output of an amplifier, but this is usually dc blocked

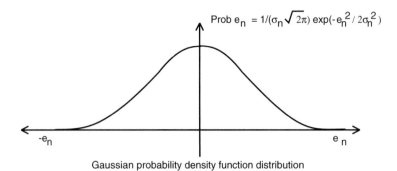

Figure 7.24 *The amplitude distribution of Gaussian white noise, showing how the larger the amplitude, the less likely it is to occur*

before application to the next stage. The noise is then, strictly speaking, no longer baseband noise, being in effect highpass filtered with some (generally low) cut-off frequency.

The value σ_n is not only the standard deviation of the noise voltage, it is also the rms value of the waveform. Whilst the peak value of a sinewave is exactly $\sqrt{2}$ times the rms value, there is no hard and fast limit in the case of noise. Some circuits have to handle a noise-like signal, as, for example, in FDFM (frequency division frequency multiplex telephony). It is necessary in such cases to design for a headroom of four or five sigma, i.e. four or more times the rms amplitude of the noise. Signal magnitudes greater than 4σ occur for less than 0.01% of the time, so although overloading will occur, it is very infrequent. Thus the peak factor for an amplifier which must handle a random noise-like signal is ×4 or 12 dB. The peak factor (i.e. peak value over rms value) for a sinewave is, as noted above, $\sqrt{2}$ or $\sqrt{3}$ dB. Thus the power handling capacity of an amplifier which must handle a random noise-like signal is 9 dB less than for a sinewave.

Other types of noise

Thermal noise can be described as Gaussian white noise. Noise in semiconductor devices approximates to a Gaussian white characteristic over a limited range. Active devices such as transistors and opamps depart from this at both ends of the spectrum. At low frequencies, the noise increases relative to that at mid-frequencies. Its level eventually becomes inversely proportional to frequency, below the '$1/f$ noise corner frequency'. Depending on the device, the $1/f$ corner frequency may be anything from tens of kHz down to a few Hz or less. Being out of band, $1/f$ noise is usually no problem in an rf amplifier stage. But in an oscillator, the non-linearity inherent in oscillator action results in the active device's $1/f$ noise being cross-modulated onto the oscillator's rf output, as close-in noise sidebands.

White noise (constant power per unit bandwidth) may be filtered to produce a level which is no longer independent of frequency. Pink noise is noise with an amplitude which falls with increasing frequency, at a rate of 3 dB/octave. It possesses the characteristic of constant power per octave, and may be used in audio testing. Red noise falls at 6 dB/octave, and as such matches the signal handling capacity of a delta modulator. It may be used in such a circuit to simulate voice loading, since the higher frequency (unvoiced) components of speech such as sibilants are at a relatively much lower level than the lower frequency voiced components. By analogy, noise whose level *rises* at 6 dB per octave may be described as blue noise, but I have yet to come across any practical application for it.

PRBS (pseudo-random bit sequence) generators make a convenient source of baseband noise, within certain limitations. The output approximates a white distribution up to $f_c/2\pi$, i.e. about one-sixth of the clock frequency. It actually consists of a series of discrete spectral lines, being the fundamental and harmonics of the frequency $f_f = f_c/(2^n - 1)$. Here, n is the number of stages in the shift register (assumed large), and the feedback is arranged to produce a maximal length pseudo-random sequence (which repeats after $2^n - 1$ clock cycles). But whilst approximately white from f_f to $f_c/2\pi$, the output is not Gaussian, consisting of a pseudo-random sequence of logic 0s and 1s. It can be rendered approximately Gaussian by passing it through a single-pole lowpass filter with a cut-off frequency of f_c/n. Now, due to the heavy filtering, the rarer longer runs of 0s and 1s have a chance to build up to larger peaks, compared with the lower amplitude of successive reversals.

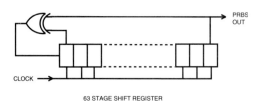

Figure 7.25 *Clocked at around 10 MHz, the pseudo-random bit stream from a 63 stage PRBS generator repeats every 32 000 years*

Figure 7.25 illustrates a baseband noise generator using a PRBS. The pseudo-random sequence of 0s and 1s that it generates will repeat after $(2^{63} - 1) = 9.223 \ldots 10^{18}$ clock cycles. If it is clocked at 9.223 MHz, the sequence of 0s and 1s will repeat after some 10^{12} seconds, or about every 32 000 years. With 10^{12} discrete spectral lines in each 1 Hz of bandwidth, it clearly represents a very good approximation to the continuous spectrum of white noise (up to $F_{clock}/2\pi$ or about 1.5 MHz). For a Gaussian distribution, it should be lowpass filtered with a cut-off frequency of $F_{clock}/63$ or less, say 100 kHz. Clearly, as an audio frequency noise generator, a 63 stage shift register is wild overkill. However, it is one of the shift register lengths where a $2^n - 1$ maximal length sequence can be obtained using a single EXOR (exclusive OR) gate connected to the appropriate tappings (in this case, stages 1 and 63). Certain other lengths share this property, which results from the describing polynomial having only three non-zero terms – a trinomial.

Reference 1 describes an audio frequency noise generator using a more modest shift register of 31 stages. Suitable inputs to the EXOR gate to achieve a $2^{31} - 1$ maximal length sequence of 2 147 483 646 clock cycles are taken from stage 13 and the last stage. Clocked at a modest 220 kHz, the pattern repeats after about 2.7 hours. A higher clock frequency would be needed if audio frequency Gaussian noise –

white up to 20 kHz – was wanted. But this design was for a source of pink noise only, the pink noise filter ensuring a near-Gaussian distribution. Actually, two filters were used, providing two output channels. These could either be from the same sequence of 0s and 1s in the same phase ('mono' mode), or one with the sequence inverted ('inverse polarity' mode), or in 'stereo' mode. In the latter case, an additional EXOR gate is used to derive a time-shifted version of the sequence, which is thus, for practical purposes, uncorrelated with the other channel. The necessary power supply need consist of nothing more than a 9 V 6F22 style (e.g. PP3) layer type battery, plus a decoupling capacitor. The circuit is reproduced here as Figure 7.26.

Where a simple single-channel source of audio noise is required, there is little to beat that handy chip, the *MM5437*, from National Semiconductor. This was featured some while ago in an article in *Electronics World*, Ref. 2. This 8 pin plastic DIL device incorporates a 23 stage shift register and requires just a 5 V supply to give a white noise (pseudo-random bitstream) output, using its own internal clock generator. Alternatively, an external clock may be used, and the

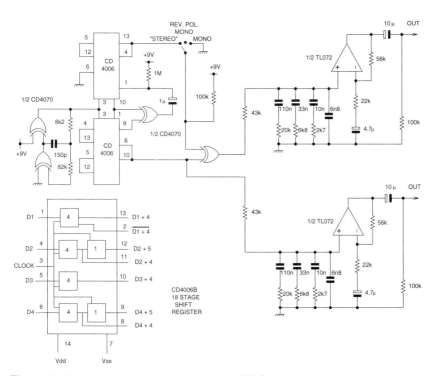

Figure 7.26 *31 stages are enough, in this PRBS generator, which provides two pink noise outputs which are effectively uncorrelated*

addition of a single-pole lowpass filter – one resistor and one capacitor – gives you noise with an approximately Gaussian distribution.

Narrowband noise

Narrowband noise may be defined as noise covering much less than one octave. Relative to a centre frequency F_c, assume that it extends over the range $-\delta F$ to $+\delta F$. Then if $2\delta F < F_c/10$, it may be considered as narrowband noise.

Narrowband noise is of particular interest to the radio engineer, as the signal presented to a receiver's detector (frequency discriminator, phase detector or whatever) will be accompanied by only that bandwidth of noise that can pass through the IF filter stage(s). Narrowband noise (thus defined) has interesting properties, since unlike baseband noise, it is not a 'real' signal. All of the information about a real signal can be conveyed on a single circuit – a single wire (plus an earth return, of course). As narrowband noise is a complex signal, it can only be completely described (i.e. in both amplitude and phase) by considering both of two separate components: in-phase and quadrature.

Figure 7.27 shows a set-up for producing a narrowband of noise, 2 kHz wide, centred on 10 MHz. Assuming the mixer is perfectly balanced, there is no component of the 10 MHz carrier frequency present in the output. The noise power per unit bandwidth is constant over the range 9.999 MHz–10.001 MHz, with a roll-off above and below those frequencies identical to the roll-off of the 1 kHz baseband filter used to define the width of the baseband noise.

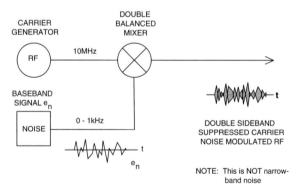

Figure 7.27 *This circuit produces DSBSC modulated noise, which is not the same thing as narrowband noise*

However, the resultant narrowband noise bears no resemblance to naturally occurring narrowband noise. As Figure 7.27 shows, every time the baseband noise waveform crosses the zero voltage axis, there is a zero in the amplitude of the 10 MHz-centred narrowband noise. Between these zeros, or cusps of the rf, the phase of the signal is coherent, whilst at each cusp there is an instantaneous phase reversal of exactly 180°.

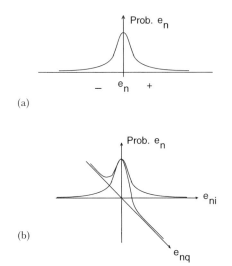

Figure 7.28 (a) Voltage probability distribution of a real signal (same as Figure 7.24). (b) Voltage probability distribution of the in-phase and quadrature components of a narrowband noise (a complex signal)

Figure 7.28(a), which has appeared earlier as Figure 7.24, describes in statistical terms the distribution of the baseband noise, but it does not describe the distribution of the rms value of the narrowband rf noise. To illustrate true narrowband noise, imagine a second mixer, whose output is added to that of the mixer output in Figure 7.27. Further, that the second mixer is fed from the same rf generator, but with 10 MHz shifted in phase by 90°. Also, that the 0–1 kHz baseband noise fed to the second mixer comes from an entirely different source, having zero correlation with the noise fed to the first mixer. The distributions of the in-phase and quadrature noise sources are sketched in three dimensions in Figure 7.28(b). Now, instead of the phase of the 10 MHz noise being either zero or 180°, it can take any value over 0–360°, with equal probability. The fact that the two baseband noise sources were supposed uncorrelated leads to an intriguing paradox.

Although clearly the most likely value of the baseband voltage at any instant is zero, voltages just either side are almost as likely, only becoming very unlikely at plus or minus two or three sigma or more. But because the baseband noise waveforms have zero correlation, the likelihood of one being *exactly* zero at the *same instant* as the other passes through zero is *vanishingly small*, i.e. zero. Consequently, there are dips in the envelope of the noise, and these are more cusp-like the deeper they are (but a complete dropout has zero probability),

312 Analog circuits cookbook

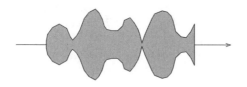

This is narrowband noise

Figure 7.29 *Illustrating the envelope of narrowband noise (see text)*

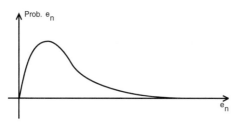

Rayleigh distribution of rms amplitude of narrowband noise

Figure 7.30 *The Rayleigh distribution*

as illustrated in Figure 7.29. This waveform simulates exactly true narrowband random noise, the rms value R of which exhibits a 'Rayleigh' distribution, sketched approximately in Figure 7.30. The Rayleigh probability $p(R)$ is given by:

$$p(R) = \frac{R}{\sigma^2} \exp\left(\frac{-R^2}{2\sigma^2}\right)$$

Unlike baseband noise where the rms value is σ, the rms value of narrowband noise with its Rayleigh distribution is $\sqrt{2}\sigma$.

The noisy signal

A noisy signal may be considered, in the simplest state, to be a steady state CW signal plus narrowband noise. The CW could be, for example, the mark tone of an FSK signal. As the level of the CW relative to noise is increased, from a signal-to-noise ratio of minus infinity dB, the Rayleigh distribution starts to change. Very low values become less and less likely, whilst as the SNR becomes positive and then large, the distribution narrows down towards the amplitude of the CW, as illustrated in Figure 7.31 (a rough representation, not to scale). This is called a Ricean distribution. It describes the signal at the back end of a receiver's IF strip, just before the detector. The noise accompanying the signal may have been picked up by the antenna, or it may be the front-end noise of the receiver. But either way, it will have been band limited by the selectivity built into the IF strip.

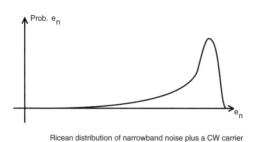

Ricean distribution of narrowband noise plus a CW carrier

Figure 7.31 *The Ricean distribution*

Stationary, or not?

The Ricean distribution, like the Rayleigh, assumes the noise in question is 'stationary'. All the types of noise considered so far have been stationary, that is to say their characteristics have been continuous, unvarying, their statistics independent of time. Certain types of noise are non-stationary, the most obvious example being impulsive noise. This is typically due to a number of causes, including vehicle ignition systems, electrical machinery and switches, and meteorological electrical activity. For signals having a large amount of redundancy, e.g. speech, impulsive noise is mainly just a nuisance, but in a data link carrying digital information, its effect can be devastating. Such links therefore usually incorporate at least an *error detection* algorithm. A parity bit per character ('8-bit ASCII') is the simplest form, but this will not detect a double error, and so more complicated schemes such as Reed–Soloman, etc. – often incorporating *error correction* in addition – are usually required.

Carrier noise

In a wireless communications link employing phase modulation – e.g. DPSK, MSK or whatever – various sources of noise contribute to the final BER (bit error rate) achieved. The most obvious is noise picked up by the antenna, or due to the noise figure of the receiver's input stage(s). Another is the phase noise of the carrier on to which the transmitter modulates the data, and yet another the phase noise of the local oscillator in the receiver. Consequently, however large the received signal, there is usually a small but finite irreducible BER, hopefully – in a well-designed system – much less than 1 in 10^4 and often of the order of 1 in 10^7. Figure 7.32 shows (much exaggerated) how the output of an oscillator exhibits random noise sidebands, resulting in both residual noise AM and residual noise FM. Frequently, the amplitude of the AM noise sidebands is negligibly small relative to the FM noise sidebands. But in any case they are irrelevant in an FM or PM link, were a limiting IF strip is used.

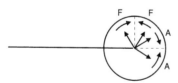

Sine wave with AM and FM noise sidebands (A, F), grossly exaggerated

Figure 7.32 *Illustrating an ideal noise free CW carrier, with (at its tip) AM and FM noise sidebands*

Figure 7.33 sketches a typical oscillator output, and indicates that beyond a certain distance from the carrier, there is a flat noise 'floor'.

314   Analog circuits cookbook

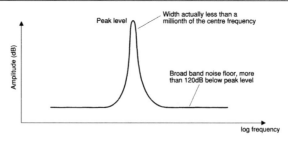

Figure 7.33 *The resulting spectrum looks something like this*

In a high quality crystal oscillator, this may be at −140 dBc (140 dB below the carrier power), from as close in as 10 Hz offset. In an LC oscillator, the noise floor may be only 90 dB down or even less, with this level not being reached until an offset of perhaps as much as 100 kHz.

Figure 7.34 shows how noise sideband power is defined. It is the level (measured in a 1 Hz bandwidth) relative to the total carrier power, as a function of the offset from the centre frequency f_o. Figure 7.35 shows in more detail the various components of sideband noise. In practice, the various stages are often not discernibly distinct, tending to run into each other.

The carrier voltage, complete with the noise modulation, is described by the expression

$$v(t) = V_s \cos[2\pi f_o t + \Delta\phi(t)] \tag{7.8}$$

where V_s is the peak value of the carrier (this expression assumes that the AM noise is negligible compared to the phase noise). Function $\Delta\phi$ of t is the randomly fluctuating phase noise term.

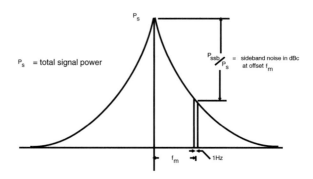

Figure 7.34 *Defining sideband noise $\mathscr{L}(f_m)$*

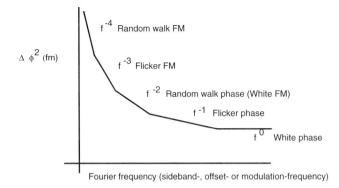

Figure 7.35 *Showing the various mechanisms responsible for the observed noise sidebands*

Is it noise?

Or is there some CW signal there? In a few specialised applications it is important to know whether, e.g., the IF signal in a surveillance receiver is pure noise, or whether there is also a weak CW signal lurking in there somewhere. Assume the IF is at $Fo = (\omega_o/2\pi)$ Hz, and that the bandwidth B ($=2\omega_b/2\pi$) Hz centred on ω_o. Assume further that $\omega_o \gg \omega_b$ and that the filter shape approximates a rectangular ('brickwall') shape. Then the variance of the number of zero crossings N of the hard-limited signal in a sample time T (seconds) – for the case where $BT \gg 1$ – is given by:

$$VAR[N(T)] \cong 0.62 \frac{2\omega_b T}{2\pi} \qquad (7.9)$$

so the standard deviation is

$$\sigma[N(T)] \cong 0.782(BT)^{\frac{1}{2}} \qquad (7.10)$$

An important result that can be found in Ref. 3.

In the event that the standard deviation over a number of sample periods each of T seconds is significantly less than this, then a CW signal must be present. Ref. 3 also gives an expression for $VAR[N(T)]$ for the Ricean case, the sum of an unmodulated carrier plus narrow band noise.

References

1. Muller, B. (1984) A stereo noisemaker. *Speaker Builder*, April.
2. Hickman, I. (1992) Making a right white noise. *Wireless World*, March, pp. 256, 257, with four useful references to articles giving

the feedback connections for maximal length sequences, for various length shift registers. Reproduced in Hickman, I. (1995) *The Analog Circuits Cookbook.* Butterworth-Heinemann, ISBN 0-7506-2002-1.
3. Roberts, J.H. (1977) *Angle Modulation.* Peter Peregrinus Ltd, ISBN 0 901223 95 6.

> **Oscillator phase noise**
>
> Oscillator purity is an increasingly important factor for designers of communications equipment. This article investigates phase noise in rf oscillators, and highlights one important factor – whether the maintaining transistor is allowed to bottom or not.

Understanding phase noise

Introduction

Modern wireless communication often uses one or other of the various types of digital modulation. The earlier, simpler forms, such as basic DPSK (binary phase shift keying), are relatively robust, requiring only a modest signal-to-noise ratio at the receiver to guarantee successful reception. But shortage of spectrum space spurred the search for greater bandwidth efficiency. This led first to the development of variations on the theme of QPSK (quadrature phase shift keying), which conveys two bits of information per signal element or 'symbol'. Later, more exotic forms, such as 16PSK, 64APK and even 256APK appeared, carrying respectively four, six and eight bits per symbol.

At the receiver, the demodulator must effectively measure the phase difference between successive symbols. This starts out, at the transmitter, as 0 or 180° – in the case of asymmetrical DPSK – or only ±90 degrees in the symmetrical form. But on reception, the effect of noise and interference is to erode the available phase margin, possibly leading to bit errors. With QPSK the phase change between symbols is 0, ±90 or 180° (asymmetrical form), or +45, +135, –45 or –135 in the symmetrical case ('$\pi/4$ QPSK'). So a higher signal-to-noise ratio at the receiver is required for the same BER (bit error rate).

With the advanced forms of modulation mentioned earlier, the phase change from one symbol to the next may be only 22.5° or even less, so clearly an even greater signal-to-noise ratio is required for an acceptable BER.

Noise in the receiver

Atmospheric noise and interference are not the only problems a digital data receiver faces. Whilst an HF receiver with a reasonably efficient aerial is likely to be 'externally noise limited', at VHF and even more so at UHF and microwaves, external noise is so low that reception will usually be limited by the receiver's own noise. One usually thinks, in this context, of input stage noise. But in the reception of digital phase modulation, an important contribution to the factors eroding the essential phase discrimination, on which a low BER depends, is the phase noise of the local oscillator.

Ideally, an oscillator produces an isolated spectral line, with zero energy output at any other frequency. Of course, there will be some harmonic content, but this is usually unimportant in a well-designed receiver. Much more troublesome is energy at frequencies immediately adjacent to the oscillator output. This takes the form of noise sidebands, which can be quite large at very small offsets from the oscillator frequency, falling off at greater offsets, until at frequencies well removed from the carrier, their level bottoms out at the oscillator's far-out noise floor.

Why phase noise is important

The sidebands consist of a mixture of amplitude noise and phase noise. In a receiver local oscillator application, the amplitude noise sidebands are usually unimportant, since the local oscillator output is applied to the mixer at a high level: the LO input of the mixer thus operates in a heavily compressed mode. So minor level changes – even of a dB or so – would have negligible effect. But the LO phase noise is quite a different story. The IF signal reflects the phase difference between the rf signal input and the LO drive waveform. Thus LO phase noise adds linearly to phase disturbances of the wanted signal. These include noise, interference and multipath suffered in the over-the-air path, and front-end noise due to a marginal signal level.

The over-the-air path is outside the receiver designer's control; he can only concentrate on the other factors, of which – in a digital data receiver – oscillator phase noise is a major component.

Phase noise of the LO

A receiver's local oscillator may, in special cases of fixed frequency operation, be a crystal oscillator. Such an oscillator is characterised by extremely low levels of sideband noise – which is usually denoted by

$\mathscr{L}(f_m)$ and defined as the noise power in a 1 Hz bandwidth at an offset of f_m. But usually the LO will be an LC oscillator, and these exhibit a higher level of sideband noise, extending out much further on either side. To highlight the difference, note that a good crystal oscillator may show a level of sideband noise, $\mathscr{L}(f_m)$, which is already down to −140 dBc at only 10 Hz offset from the carrier. By contrast, a commercially advertised varactor-tuned VCO module, covering the range 100–200 MHz, claims a typical $\mathscr{L}(f_m)$ of −105 dBc at 10 kHz offset, and around −120 dBc at 100 kHz offset.

Where the LC oscillator forms the VCO in a phase-locked loop, its sideband phase noise within the loop bandwidth will be reduced by the loop negative feedback, but outside the loop bandwidth will return to the level it would be were the VCO running open loop. Clearly, even given a degree of phase-noise clean-up by the loop, one is better off starting out with a low phase-noise oscillator in the first place. A facility for measuring the phase noise of an oscillator is therefore an important item in any rf development lab, and can involve some very expensive equipment. I was therefore interested in an article which described such a measurement system using only standard lab instruments plus some inexpensive bits of rf kit, Ref. 1. The basic arrangement is shown in Figure 7.36.

To B or not to B(ottom)?

I wanted to try and measure the phase noise of an oscillator, in order to settle a question which has interested me for some time. Namely, is there an advantage in designing an LC oscillator in such a way that the transistor does not bottom at the negative-going peaks of the waveform? In fact, many LC oscillator designs do result in the transistor bottoming, and indeed this can be quite difficult to avoid in an oscillator with a wide tuning range such as a three-to-one frequency ratio, given production spreads in transistor characteristics. The effects of bottoming in an rf oscillator had been explored in an earlier article, Ref. 2, but equipment to measure phase noise was not available to me at that time.

An LC oscillator was therefore built up, operation at around 10 MHz rather than VHF being chosen, as more readily manageable for measurement purposes. This, together with the other items needed for the Figure 7.36 type set-up, is shown in detail in Figure 7.37. The tank circuit inductor L_1 was a Coilcraft *SLOT-TEN-1-03* unshielded inductor with a carbonyl E core, having a quoted nominal inductance of 2.2 microhenries and Q of 56 at 7.9 MHz. A Colpitts oscillator circuit was chosen, as the inductor was untapped, arranged so that the transistor could be operated with the emitter connected

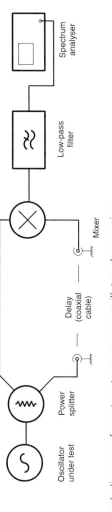

Figure 7.36 Block diagram of a set-up to measure oscillator phase noise

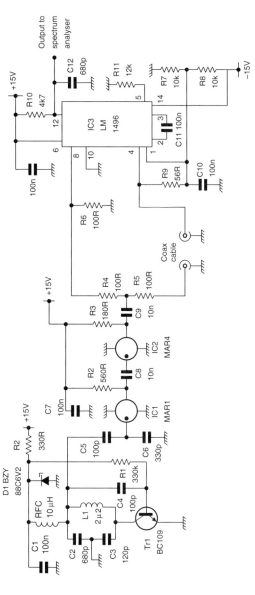

Figure 7.37 Circuit diagram of an experimental set-up to measure oscillator phase noise

directly to circuit ground. To minimise loading and maintain a reasonably high working Q, the output was taken from the base end of the tank circuit. This is a much lower impedance point than the collector end, and loading was further reduced by using a capacitive divider, C_5 and C_6, to buffer the 50 Ω input of IC_1. Together with IC_2, IC_1 provides a total gain of 26.7 dB nominal, providing a level of –8 dBm into 50 Ω at the coaxial socket connected to R_5.

The frequency discriminator

The output of IC_2 (which sees a 50 Ω load approximately) is applied to the LO port of an active double balanced mixer, IC_3, an *LM1496*. Figure 7.38(a) shows the internal circuit of IC_3. The 'carrier' or LO is applied between pins 8 and 10, to four transistors connected in an arrangement often referred to as a Gilbert Cell. The signal input is applied between pins 1 and 4, the signal being steered in phase or in antiphase to the outputs at pins 6 and 12 (note the pin numbers quoted refer to the DIP packaged version of the *LM1496*). The transconductance of the signal long-tailed pair is set by the value of a resistor connected between pins 2 and 3. The magnitude of the tail currents is set by the current injected into the bias port, pin 5.

Figure 7.38(b) shows how the output at pin 12 is at its maximum positive level if the LO and signal are in phase, is at zero (relative to its level in the absence of a signal input) when they are in quadrature, and at maximum negative level when in antiphase. If pins 2 and 3 are shorted, so that both signal and LO ports are overdriven – equivalent to squarewave drive in each case – the input phase to output voltage characteristic is linear – Figure 7.38(b), right-hand side. If the signal port is operated in a linear manner, the characteristic is cosinusoidal, also shown in Figure 7.38(b).

In Figure 7.37, the signal is applied to the signal input port via R_5 and a length of coaxial cable. The latter provides a fixed time delay, independent of frequency. Therefore if the oscillator frequency is varied, the electrical length of the cable varies, and so the phase of the signal applied to pin 4 of IC_3 will vary. So although IC_3 is a phase sensitive detector, in conjunction with the delay cable it forms a frequency discriminator.

The delay was provided by a reel of miniature polythene insulated coaxial cable, unearthed from my stock of handy bits and pieces. This coax had a silver on copper on steel inner, and might or might not have been UR94. Monitoring pin 4 of IC_3 with one 'scope probe and the junction of R_4 and R_5 with the other, the waveforms were found to be in quadrature at 10.377 MHz and in antiphase at 9.726 MHz. From these results, and assuming the velocity of propagation in the

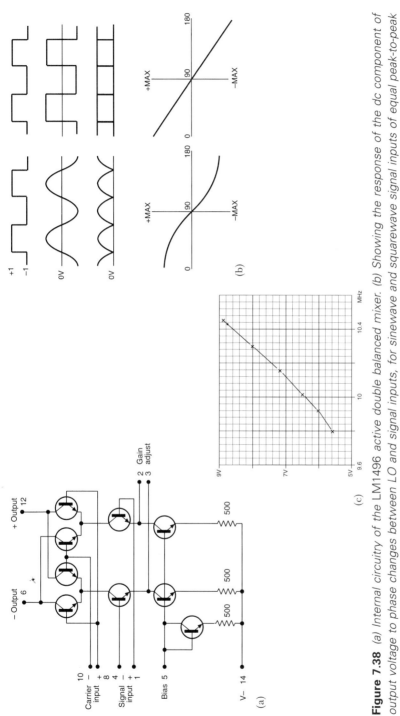

Figure 7.38 (a) Internal circuitry of the LM1496 active double balanced mixer. (b) Showing the response of the dc component of output voltage to phase changes between LO and signal inputs, for sinewave and squarewave signal inputs of equal peak-to-peak voltage, assuming linear operation of the signal port. (c) Showing the response of the Figure 7.36 frequency discriminator, as implemented in Figure 7.37. The black dot shows the measured centre frequency response, crosses show other measured points

cable is two-thirds that in free space, some simple algebra gives the length of coax as 14.95 quarters of a wavelength at 10.377 MHz, say 3¾ wavelengths, allowing for experimental error. Thus T_d is 361 ns and the physical length of the cable turns out to be 72.3 m. I took this figure on trust, rather than unreeling the cable to find out!

Frequency discriminator sensitivity

Maximum sensitivity is ensured by C_{11} which provides an ac short between pins 2 and 3 in Figure 7.37. As the dc resistance between these pins is infinite, in the absence of a signal input, the output sits at the midpoint of the characteristic, despite any small input offset voltage that there might be between pins 1 and 4.

By varying the tuning with the core of L_1, measuring the frequency with a digital frequency meter and the output level at pin 12 of IC_3 with a DVM, the frequency discriminator characteristic was measured. This is shown plotted in Figure 7.38(c). Due to the limited available tuning range, for the most part, only one side of the characteristic could be plotted, as shown. The considerable length of coaxial cable used achieved a high sensitivity in the frequency discriminator, but introduced some inevitable attenuation. Consequently the signal voltage swing available at pin 4 was less than the LO input at pin 8, the attenuation in the cable being some 7 dB. The result is that the discriminator characteristic is intermediate between those shown in Figure 7.38(b). Over the central linear portion, the characteristic sensitivity is 164 kHz/V or 6.09 μV/Hz.

The measured results

The output of the frequency discriminator, at pin 12 of IC_3, was connected to an *HP3580A LF* spectrum analyser, via the lowpass filter shown in Figure 7.36. Figure 7.37 shows that the filter consisted simply of the 4K7 Ω phase detector output resistor R_{10}, in conjunction with some 800 pF or so. This consisted of C_{12} plus about 100 pF due to a screened input lead and the analyser's input capacitance. The cut-off frequency of this filter is a little over 40 kHz, well clear of my range of interest, which was in noise sidebands up to 5 kHz.

First of all, to establish a measurement noise floor, a spectrum analyser sweep from 0 to 5 kHz was recorded with the power supplies switched off, Figure 7.39(a), lower trace. This shows a measurement noise floor of about 80 dB below a top-of-screen reference level of −60 dBV, or some −140 dBV. At this level, it is difficult to avoid some response from supply rail residual hum, visible as 100 Hz and harmonics thereof at the left-hand side of the trace.

Next, the circuit was powered up, but with the coax cable disconnected. Figure 7.39(b) upper trace, 5 V/div., 0 V at centreline, shows the standing voltage at the frequency discriminator output, IC_3 pin 12. This was +8.75 V – corresponding to the discriminator centre frequency. The coax was then reconnected and the lower trace (50 mV/div., 20 ns/div. ac coupled) shows the delayed signal applied to IC_3 pin 4. Some modulation of the trace is visible, but this was still there when the supplies were turned off – it turned out to be pick-up of the local FM radio station. As the frequency is unrelated to the LO waveform at IC_3 pin 8, it will not affect the result and can be safely ignored.

With the coaxial cable reconnected, the frequency was adjusted to 10.377 MHz, by means of the core in L_1. At this frequency the signal input at pin 4 of IC_3 was in quadrature to the LO input at pin 8, corresponding to zero deviation from the discriminator's centre frequency. The oscillator's phase noise sidebands (on both sides of the carrier) are translated by the frequency discriminator to baseband – from 0 Hz upwards. The result is displayed in Figure 7.39(a), upper trace. This is over 30 dB clear of the measurement noise floor, due to the high system sensitivity ensured by the generous length of coax employed.

The corresponding value of $\mathcal{L}(f_m)$ at 2.5 kHz offset was calculated as shown in the box on p. 327. The result seems plausible, even if only an approximation. However, for the purposes of comparing phase noise with the transistor bottoming, or not bottoming, comparative measurements suffice, and proved revealing, as shown below.

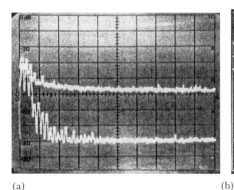
(a)

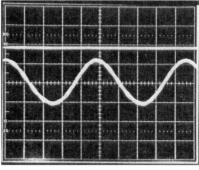
(b)

Figure 7.39 *(a) Spectrum analyser sweep, 0–5 kHz, reference level (top of screen) −60 dB, 10 dB/div. vertical, IF bandwidth 30 Hz, smoothing maximum, 100 seconds per division sweep speed. Lower trace, with + and −15 V supplies off. Upper trace, supplies on, circuit as in Figure 7.37. (b) Oscilloscope traces; horizontal, 20 ns/div. Upper trace, IC_3 pin 12, 5 V/div., 0 V at centreline, with coax cable disconnected. Lower trace, IC_3 pin 4, 50 mV/div. ac coupled, coax cable connected*

324 *Analog circuits cookbook*

I needed to know whether the oscillator was bottoming or not. An *HP8558B* spectrum analyser was used to sample the output at the base end of L_1. To avoid excessive loading of the circuit, the 50 Ω coax lead to the spectrum analyser was connected via a 4K7 resistor.

Figure 7.40(a) shows the spectrum of the oscillator, with settings of 10 dB/div. vertical, reference level –10 dBm, 5 MHz/div. horizontal, 30 kHz IF bandwidth, video filter on maximum. The illustration is a double exposure, showing the output of the circuit as in Figure 7.37 (0 Hz marker at extreme left), with the fundamental at just over 10 MHz, its second harmonic nearly 30 dB down, with the higher harmonics much lower – lost in the measurement noise floor. The second trace, with increased Tr_1 base current (offset half a division to the right), shows a larger fundamental and prominent third and fourth harmonics in addition to the second.

The second trace is the result of connecting a 56K resistor in parallel with R_1. Thus the base current was increased by a factor of over six, whilst the output amplitude increased only by some 8 dB or ×2.5. This, together with the marked level of higher harmonics, shows that with the additional base current the circuit was bottoming, but without it was not.

Figure 7.40(b) shows (upper trace) the 0–5 kHz baseband spectrum, with the increased base current, resulting in the transistor bottoming. The lower trace is a repeat of the upper trace in Figure

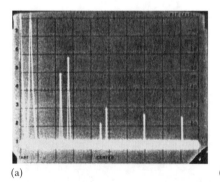
(a)

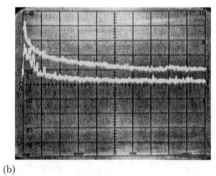
(b)

Figure 7.40 *(a) Spectrum of the oscillator, with settings of 10 dB/div. vertical, reference level –10 dBm, 5 MHz/div. horizontal, 30 kHz IF bandwidth, video filter on maximum. Double exposure. Circuit as in Figure 7.37 (0 Hz marker at extreme left) shows the fundamental at just over 10 MHz and its second harmonic nearly 30 dB down. Trace with increased Tr$_1$ base current (offset half a division to the right) shows larger fundamental and prominent third and fourth harmonics. (b) Upper trace, 0–5 kHz baseband spectrum, with increased Tr$_1$ base current, transistor bottoming. Lower trace, repeat of the upper trace in Figure 7.39(a), for comparison. Both with same settings as Figure 7.39(a)*

7.39(a), for comparison. Both traces were recorded with the same settings as Figure 7.39(a). For this test, care was taken that the signal applied to the frequency discriminator was the same as without the increased base current. To this end, after adding the 56K resistor in parallel with R_1, the 100 pF capacitor C_5 was replaced by a 5–65 pF trimmer. This was adjusted to give the same amplitude inputs at the LO and signal ports of IC_3 as previously. The resultant small shift in oscillator frequency, due to the slightly reduced loading on the tank circuit, was removed by readjusting the core of L_1.

Conclusions

It can be seen from Figure 7.40 that in the range above 2.5 kHz offset, the magnitude of the phase noise relative to the carrier is nearly 10 dB lower when the transistor is not bottoming than when it is. Note particularly, that the gap widens at lower offsets. This is presumably because bottoming involves higher order non-linearities, resulting in the transistor's $1/f$ noise, cross-modulated onto the carrier, effectively extending further out into each sideband.

By 5 kHz, the noise (as measured with a frequency discriminator) has clearly flattened out. This corresponds to phase noise falling at 6 dB/octave of offset frequency, or the f^{-1} region of phase noise, which continues until the far-out noise floor is reached. At smaller and smaller offsets, the slope becomes greater, f^{-2}, f^{-3} and at very small offsets f^{-4}. This tendency is visible in both traces in Figure 7.39(a), though setting in at a higher frequency when the transistor is bottoming. As the offset reduces to zero, the amplitude increases, up to the value of the carrier output. The trace in Figure 7.39(a) does not show this below 5 Hz, as this is the low frequency limit of the *HP3580A* spectrum analyser. In any case, the output due to the carrier itself is (near) zero, since the LO and signal inputs are in quadrature.

So when an oscillator with low phase noise is required, a circuit design should be selected which avoids bottoming of the collector. This can be achieved in a number of ways, for instance using a 'long tail' to define the emitter current, Figure 7.41(a). Where a large tuning range is involved, it may be advantageous to vary the tail current. Assuming capacitive tuning, the dynamic resistance of the tank circuit will increase with frequency. So to maintain a constant amplitude of oscillation, the tail current should be varied inversely as the oscillator frequency.

Of course, even when not bottoming, the transistor is still operating non-linearly, the collector current being cut off for part of each cycle. If amplitude control could be implemented independently

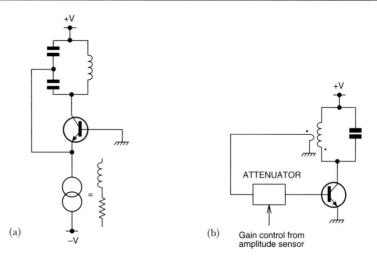

Figure 7.41 (a) Defining the transistor's collector current. By means of a long tail as here is just one of many ways. The resistor may be replaced by the output of a DAC, permitting adjustment of the tail current under program control. (b) Separating the amplitude control mechanism from the oscillator should permit operation of the transistor in a linear regime. This should result in much reduced phase noise sidebands, by preventing the transistor's 1/f noise cross-modulating onto the carrier

of the transistor, as indicated in Figure 7.41(b), it should be possible to operate the transistor entirely in a linear mode, preventing the cross-modulation of its $1/f$ noise onto the carrier output. An interesting possibility which I pursued in a later article in *Electronics World*, Ref. 3. Doubtless this has been done many times already, but I don't recall having seen the results published elsewhere.

An alternative to Figure 7.41(b) would be to use a VGA (variable gain amplifier) as the maintaining amplifier. A suitable candidate would seem to be the recently announced *CLC5523* (from National Semiconductor, with a 250 MHz bandwidth at 135 mW power consumption), of which I am trying to obtain a sample.

In Figure 7.40(b) (lower trace), the measured level of sideband noise at 2.5 kHz offset from carrier, with the circuit of Figure 7.37, is −108 dBV in a 30 Hz measurement bandwidth. To work out $\mathscr{L}(f_m)$, the value in a 1 Hz bandwidth is needed. The analyser's IF filters consist of five synchronously tuned crystal filter stages, providing a Gaussian response. This characteristic is optimum for rapid settling to the true value of a swept signal. The noise bandwidth of such a filter is 12% greater than the actual −3 dB bandwidth. The nominal 30 Hz bandwidth is subject to a ±15% tolerance, so the actual −3 dB bandwidth was measured, using the 1 dB/div. scale. This turned out to be 27 Hz, giving a noise bandwidth of 30 Hz, as near as makes no odds. Thus the level of −108 dBV in 30 Hz translates to −123 dB in a 1 Hz bandwidth. This represents the sum of the noise energy in both upper and lower sidebands, giving a figure of −126 dBV or 0.5 μV for the single sideband noise.

Given the measured sensitivity of the frequency discriminator of 6.1 μV/Hz (see above), the rms frequency deviation f_d is 0.082 Hz. For sinewave modulation at a frequency f_m, the modulation index $m = f_d/f_m$ equals the peak phase deviation in radians. Now $0.082/2500 = 3.3 \cdot 10^{-5}$ radians, and for such a small phase deviation, only the first-order FM sidebands are significant. So if the modulating frequency f_m were a 2.5 kHz sinewave rather than narrowband noise, the first order sidebands would each be $(3.3 \cdot 10^{-5})/2$ in amplitude relative to the carrier, since for small angles, $\arctan \theta = \tan \theta = \sin \theta = \theta$, with negligible error. So the sinewave single sideband amplitude would be simply $20 \log(1.65 \cdot 10^{-5})$ relative to the carrier, or −96 dBc, and this may be taken as a first-order approximation to the value of $\mathscr{L}(f_m)$ at 2.5 kHz offset, for the circuit of Figure 7.37.

References

1. Suter, W.A. (1995) Phase noise measurement for under $250. *RF Design*, September, pp. 60–69.
2. Hickman, I. (1994) The ins and outs of oscillators. *Electronics World*, July, pp. 586–589.
3. Hickman, I. (1997) Killing noise. *Electronics World*, October, pp. 817–823.

Index

ADC (analog to digital converter, A/D, A-to-D), 43, 105, 109, 126, 205
Address code, 298
Admittance:
  output, 155
AGC (automatic gain control), 57, 142, 144, 279, 286
A_L (inductance/turn^2), 267
Alias, 16, 105
Amplifier:
  buffer, 35, 130
    unity gain, 78, 97
    FET, 138
  cascode, 155
  differential, 146
  distributed, 16
  inverting, 165, 199
  logarithmic, 10–16, 166
  IF:
    swept gain, 10
  instrumentation, 99, 100
  isolation, 198, 203
  rf, 261
  summing, 163
  transconductance, 97
  transimpedance, 214
  transresistance, 205
AM, 142, 170, 269
Amplitude modulation *see* AM
Antenna, 298

ASIC *see* IC
Asymptote, 87
Attenuation, 31, 85, 168
Attenuator, 14, 146, 202
  input, 139
Avalanche, 17

Ballast:
  choke, 217
  electronic, 218
  high frequency, 217
Bandwidth:
  IF, 38, 120, 135, 137, 169, 173, 174, 272
  noise, 327
Batteries, 228–241
  layer type PP3, 6, 9, 228, 239
  NICAD, 228, 240
  zinc/carbon (Leclanché), 229, 239
BBD (bucket brigade device), 79–83
BER (bit error rate), 313, 317
Birdie marker, 9
Bootstrapping, 110–116, 231
Bottoming, 318, 324, 325
Bridge, 92, 99, 103, 188, 251
  Wien, 63, 69
Buffer *see* Amplifier

Cable, 130, 139
  coaxial, 22, 23, 132, 138, 139, 320, 322

Camcorder, 94–98
Capacitance ~or, 26, 39, 152, 155, 157, 159
  blocking, 144
  bypass:
    RF, 144
  decoupling, 156
  electolytic, 159
  input, 128
  negative, 1–9
  polystyrene, 35
Carrier, 17, 21, 320
  amplitude modulated (AM), 142
  suppression, 285
Cassette, 97
Cell:
  AA219, 228
  C, D, 228
  Gilbert Cell, 320
  leakproof, 239
Cents, 70
Channelling, 297
Charge:
  injection, 72
Circulator, 177–190
Clipper, -ing, 68, 70
CMOS (complementary metal oxide silicon), 45, 66, 208, 232, 265, 298, 299
CMRR, 77, 100, 102, 103, 110, 203
Coax (coaxial cable) *see* Cable
Common mode rejection *see* CMRR
Comparator, 195, 253, 271
Components:
  discrete:
    active, 41
    passive, 39
  surface mount, 39–56
Conductance:
  conversion, 284
Converter F–to–V, V–to–F, 198,
Correlation, 311
Coupler, 66

CRT (cathode ray tube), 223
Crystal:
  oscillator, 314, 317
  quartz, 26
Current:
  dark, 212
  housekeeping, 230, 233, 239
  limit, 252, 262, 264
  mirror, 254
  short circuit, 256
CW (continuous wave), 269, 312, 315
Cypher, 63

D (=1/Q), 85
DAC (digital to analog converter D–A), 43, 174, 279
Darlington, 97, 117, 251, 256, 262
Data:
  acquisition, 208
  serial, 208
DDS (direct digital synthesis), 122
Delay line, 22
Demodulator, 316
Detection ~or, 100, 142–151, 286
  average-responding, 151
  crystal, 142
  edge, 194
  infinite impedance, 147
  peak-to-peak, 150
  photo, 191
  synchronouos, 286
Deviation, 273
Differentiation, 275
Diode:
  commutation, 251
  laser, 17, 213, 226, 227
  LED (light emitting ~), 198, 199, 202, 213, 217, 230, 234, 300
  infra red, 213
  silicon photo, 205–219, 214–227
  PIN (P-intrinsic-N), 199
  Schottky, 144
  thermionic, 142

variable capacitiance (varactor), 26, 163
zener, 115, 240, 256
Direct digital synthesis *see* DDS
Directivity, 178, 184
Discriminator:
frequency, 322
Dispersion, 162, 170
Dissipation, 159, 262
Distortion, 57, 60
Driver:
line, 212
DSP (digital signal processing), 104, 126, 208, 272, 279, 281
Ducking circuit, 94
Duty cycle, 296
DVM *see* Meter
Dynamic range, 104, 144, 147, 168, 208

Earth
virtual, 205
Electron, 17, 178,
EMF (electromotive force), 31
ENCU (enamelled copper), 220
Energy, 18, 22
Equipartition Law, 303
ERP (equivalent radiated power), 292, 298
ESM (electonic surveillance methods), 273
ESR (equivalent series resistance), 21

Factor:
shape, 7
FDDI, 213
Ferrite, 178
FET (J~, MOS~ VMOS~), 116, 148, 231
Fibre-optic digital data interface *see* FDDI
Filter:
active, 84–94, 104
allpass (APF), 57
anti-alias, 104, 106
bandpass (BPF), 27
Bessel (maximally flat delay), 26, 87, 93, 106, 117, 126
brickwall, 279, 315
Butterworth (maximally flat amplitude), 26, 85, 87, 90, 93, 105, 117, 119, 120, 122, 126
Caur = elliptic
Chebychev, 7, 87, 89, 90, 91, 93
crystal, 169, 327
elliptic, 26, 59, 91, 106, 117, 119, 120
equal C, 88, 89
FDNR, 26–38
FIR (finite impulse response), 93
Gaussian, 168, 327
highpass, 59, 86
Kundert, 6, 87, 88
linear phase, 93
lowpass (LPF), 59, 86, 123, 279, 322
N-path, 6–9
notch, 92, 107
post-detection, 166
Rausch, 27, 90
SAB (single active Biquad), 90–92
Sallen and Key, 7, 14, 27, 36, 84, 88, 89
second order, 85
state variable, 57,
switched capacitor (SC), 27, 105
synchronously tuned, 168, 327
termination, 31
time-continuous, 105
twin TEE, 91, 93
video, 166, 169, 324
Flip-flop, 174, 271
Foldback, 257

Free space, 10
Frequency:
    clock, 8, 27, 105, 106, 117, 163
        non-overlapping, 80, 191
    corner = cut-off
    cut-off, 85, 91, 117
FSD, 157
FSK, 269, 272, 312
Fundamental, 59, 324

Gain:
    ~bandwidth product (GBW), 45, 132, 134, 136
    differential, 101
    IF, 286
    loop, 256
        open, 100
    noise ~, 216
Gate:
    sampling, 16
    linear, 256
    logic (OR, NOR, AND, NAND, EXOR), 53, 309
Generator:
    constant current, 155
    pulse, 196
    sweep, 161, 163
    tracking, 35, 139, 186
Glitch, 17, 163, 223
Grass, 166
Ground:
    plane, 156, 182
GRP (glass-fibre reinforced plastic), 183
Guard ring, 113
Gyro:
    piezo, 73–83

Harmonics, 58, 287, 324
Heatsink, 256, 263
HF, 149
Hole, 17
Homodyne, 268–281, 281
Hum, 322

ICs (integrated circuits), 39–56
    application specific (ASIC), 56
    dual in line (DIL), 40, 88, 191
        plastic DIL (DIP), 40, 117
IGBT (insulated gate bipolar transistor), 250
Impedance, 31, 109–116
    dynamic, 3
    output, 256
Inductor ~ance, 26, 39, 100, 103, 152, 155, 157, 159
    negative, 3
    synthetic (active), 26
Insertion:
    loss, 169
Integrator ~ion, 78, 191, 192, 195, 196
Intermodulation, 57
Inverse square law, 10
Inverter, 223
Isolator, 177–190

Jitter, 20, 25

Laser *see* Diode
LED *see* Diode
Local oscillator (LO) *see* Oscillator
Limiting, 102
Lissajous figure, 172
Logamp *see* Amplifier logarithmic
Long tail, 147, 325
    LTP (long tailed pair), 80–83, 256, 281–288, 320
LPRA (low power radio association), 292

Manchester code, 298
Matrix, 211
McPherson circuit, 262
MCU *see* Microcontroller
Meter:
    level:
        audio, 13

power:
  rf, 13, 15
  volt~:
    digital (DVM, DPM), 14, 175, 257, 258, 259
    rf milli ~ ~, 147
Microcontroller, 169, 194, 195, 211
Microphone, 94
  radio ~, 289
Mike (= microphone), 97, 98
Mitochondria, 227
Mixer, 279, 310, 311, 317
  double balanced, 281–288
Modulation ~or, 143, 147, 284, 295
  amplitude, 286
  FM, 170, 268–281
  index, 172, 273
Mono, 83, 309
MOS devices (MOSFET, MOS SCR, MCT etc), 243, 246, 250
Multiplier, 279
  frequency ~, 287

Negative feedback (NFB), 29, 90, 100, 199
NEP, 218
Network analyser, 151
NICAM, 171
Noise, 145, 208, 303–327
  1/f, 109, 214, 307, 325, 326
  atmospheric, 317
  common mode, 101
  ~ equivalent power *see* NEP
  floor, 317, 322
  Gaussian (normal), 306
  impulsive, 313,
  narrow band, 310, 312
  phase, 317
  stationary, 313
Normalisation, denormalisation, 33
Notch circuit (*see also* Filter), 2
NRZ (non-return to zero data), 298
NVPD (noise voltage probability density), 306

Nyquist frequency, ~ rate, 104, 106, 126

OBW (occupied bandwidth), 295, 300
Ohm's Law, 28
Opamp (operational amplifier), 35, 45, 99–103, 253, 263, 279
  BiMOS, 99
  current feedback, 46, 179
  voltage feedback, 134
Opto, 191–227
  coupler, 202
  isolator, 198–205
  line imager, 191–198, 205
Oscillator ~ion:
  audio frequency, 57
  blocking, 220, 230
  clock, 117
  Colpitts, 318
  local (LO), 139, 175, 186, 278, 317, 320, 322
  parasitic, 158
  push-pull, 153
  voltage controlled (VCO), 79, 318
Oscilloscope, 126, 132, 137, 145
  digital storage, 172, 223
  probe, 126–141
    passive, 127
  sampling, 16, 25
OTA (operational transconductance amplifier), 279
Outphasing, 59, 60, 200, 201

Pads, 21, 187, 202
  mismatch, 1
PAM, 123
Peak factor, 307
Peaking, 132, 183
Phase:
  deviation, 327
  differential, 101
  free, 62

margin, 316
shift, 201, 214
Phono plug, ~ socket, 97
Phosphor, 218
Photo-:
  conductive, 198–204, 203
  voltaic, 198–204, 203
Piezoelectric, 75
Pixel, 191, 192
Plasma, 17, 217, 218
PMR (private mobile radio), 288
Pockel cell, 17
Polynomial, 308
Positive feedback (PFB), 64, 70
Power:
  spectral density *see* PSD
  supply, 228–267
    dual, 252
      tracking, 252
    raw, 257
    master/slave, 252
PRBS (pseudo random bit sequence), 308,
Precession, 178
Probability:
  of detection, 10
  of false alarm, 10
Programmable read only memory *see* PROM
PROM, 174, 218, 272
Propagation:
  velocity of, 321
PSD, 172
Pulse, 18
  ~amplitude modulation *see* PAM
  generator, 25
  ~ repetition frequency (prf), 20, 223
  width, 208

Q, 6–9, 26, 59, 67, 87, 91, 104, 126, 152, 153, 155, 157, 159, 320
  ~ meter, 152

Quadrature, 64, 271, 272, 310
Quartz *see* Crystal
Quieting, 300

Radar, 10
  diode/video, 145
Radio:
  FM, 83
Rank, 63
Ratio:
  mark space, 79, 121, 212
Rayleigh distribution, 312, 313
Receiver:
  paging, 271
  superhet, 269, 272, 295
Rectifier, 150
  bridge, 256
  fullwave, 13
  halfwave, 256,
Reed Soloman, 313
Reflection, 129, 178
  coefficient (symbol ρ "rho"), 177, 188
Regulator:
  voltage:
    low drop-out, 48
Rejector circuit, *see* Notch
Resistance:
  contact, 65
  dynamic, 153, 325
  input, 303
  loss, 159
  negative, 1–3, 17
    frequency dependent (FDNR), 26–38
  slope, 43
  source, 303
Resistor, 39
  load, 21
  metal film, 35
  wirewound, 263
Resolution, 208
Return loss, 178, 188
Ricean distribution, 312, 313, 315

Index

Ringing, 21, 129
Ripple, 254, 260
Risetime (falltime), 22, 24, 134, 135, 137, 212
Rms (root mean square), 307
RSSI, 12
RTL (resistor/transistor logic), 52

SAW (surface acoustic wave) device, 26, 173
SCART, 97, 98
Schmitt gate, ~ trigger, 79
SCR *see* Thyristor
Section – TEE, π, 31, 33
Sensitivity:
 tangential, 145, 147
Servo:
 bang-bang, 195
Shelf life, 238
Shift register, 191
Sidebands, 170
Signal:
 ~ to noise ratio (SNR), 208, 278, 316
 real, complex, 272, 310
 unbalanced, 101
Sinewave, 147, 199, 201
Slew rate, 97, 214
Smith chart, 152
SNR *see* Signal to noise ratio 312,
Snubber, 251
SOA (safe operting area), 251
Span, 161, 168, 169, 171, 173, 175
Spectrum:
 analyser, 139, 184
  audio frequency, 35
 monitor, 159–177
Spice model, 54
Spurious response, 166
Square law, 145
 inverse, 146
SRBP, 156, 183
SSB (single sideband), 269, 285
 noise, 327

Standard deviation, 306
Stereo, 83, 98, 309
Stopband, 85, 122
Stroboscopic effect, 172
Switch:
 T/R, 10
Sythetic resin bonded paper
  see SRBP

THD, 108, 121, 123
Tank circuit, 153, 287, 320, 325
Telemetry, 289
Tempco *see* Temperature coefficient
Temperature:
 ambient, 208
 colour, 217, 218
 ~ coefficient, 207, 210, 234
Thermistor, 57, 58, 59
Thyristor, 242
 GTO (gate turn-off), 243
Timeconstant, 112, 143, 286
Total harmonic distortion *see* THD
TPH (through-plated holes), 55
Transformer:
 variable voltage, 19
Transient, 260, 296
Transmission line, 141
Transistor:
 avalanche, 16–26
 rf, 23
 switching, 222
Triac, 242
Trigger, ~ing, 16, 25
Trinomial, 308
Tristate, 194
TTL, 232, 247
TV, 16, 223, 302
 tuner160, 176

UV (ultra violet), 218

Varactor *see* Diode
Variac, 19
VCO *see* Oscillator

Variance (square of standard
	deviation), 315
Velocity:
	propagation, 23
VHF, 147
Video, 94–98, 101
Virtual earth *see* Earth
Voltage:
	offset, 13, 78, 100, 109
	~ standing wave ratio (VSWR), 178, 187, 189

Waveform, 127
	repetitive, 16
	sawtooth, 161
	sinewave, 199
	square, 59, 67, 111, 129, 134, 163
	triangular, 172, 261

Zero
	finite (*see also* Filter – notch), 107